Please note that the previous printing included a CD-ROM, where reference in the text is made to this, please read 'Companion website'.

The material is now only available on the companion website: *http://www.elsevierdirect.com/companion.jsp?ISBN=9780750676533*

Accelerated Testing and Validation

Accelerated Testing and Validation

Testing, Engineering and Management Tools
for Lean Development

by Alex Porter

AMSTERDAM • BOSTON • HEIDELBERG • LONDON
NEW YORK • OXFORD • PARIS • SAN DIEGO
SAN FRANCISCO • SINGAPORE • SYDNEY • TOKYO

Newnes is an imprint of Elsevier

Newnes is an imprint of Elsevier
200 Wheeler Road, Burlington, MA 01803, USA
Linacre House, Jordan Hill, Oxford OX2 8DP, UK

Library of Congress Cataloging-in-Publication Data

 (Application submitted.)

British Library Cataloguing-in-Publication Data
A catalogue record for this book is available from the British Library.

ISBN-13: 978-0-7506-7653-3
ISBN-10: 0-7506-7653-1

For information on all Newnes publications
visit our website at www.newnespress.com

Transferred to Digital Printing in 2012

To my wife Theresa, whose love, patience and support made this book possible.

Contents

Preface

Often practitioners of testing, accelerated testing, reliability testing, computer modeling and other validation tools focus on the science and math of the tool: what is innovative and cutting edge (which means "cool and fun to work on"), instead of the reason for using the tools. In this sense, testing and computer modeling engineers are a lot like kids—if you give a kid a hammer, everything will need pounding; give an engineer some neat new test method, math algorithm, or computer tool and every project will need it. This is a good attribute for engineers to have; it is the excitement that brings about the exploration and development of new methods, better techniques and faster results. But for what? New methods, better techniques, more information in a shorter period of time for what? Often computer modeling and testing engineers lose sight of the reason behind what they are doing.

Testing and validation is not about conducting experiments, tests and validation demonstrations. Testing and validation is about generating key information at the correct time so that sound business and engineering decisions can be made.

The managers and quality specialists who lack this childlike fascination with testing and modeling techniques find the obsession with the science and math to be annoying. They tolerate it because they need the information that the obsessed produce with their validation tools.

This book is a cross-discipline manual for management, quality, validation, Computer Aided Engineering (CAE), and others who produce and use validation information in the product development process. Of the wide range of validation tools, this book will focus on: 1) What information is needed?; and 2) What tools can produce the information in a timely manner?

The relationship between information, time, cost and engineering decisions in the development process will be explored to provide a common dialog for making sound decisions about what information to collect, what validation tools to use and what resources to apply. Ultimately, if validation tools are selected and applied to provide the key information precisely when it is needed, the development process will not just be faster; it will be a truly efficient development process.

What's on the Companion website?

The Companion website that accompanies this book contains a host of useful material:

- Most chapters contain a directory with pictures, movies, PowerPoint® slide shows, spreadsheets and/or programs that augment and reinforce the content of the book.

The Time Value of Information

"All for want of a nail..."

Remember the old rhyme?

> *For want of a nail, a shoe was lost*
> *For want of a shoe, a horse was lost*
> *For want of a horse, a rider was lost*
> *For want of a rider, a message was lost*
> *For want of a message, a battle was lost*
> *For want of a battle, a kingdom was lost*
> *All for want of a nail.*

—George Herbert (1593-1632)

This little rhyme may be cute, and illustrate how one critical detail can ruin your whole day, but it is extremely relevant to the issue of this book and this chapter.

Without the right piece of information at the right time, the battle and the war were lost. The right information is critical to making any plan successful. The reason is that all plans have decisions that must be made at different times in order to know how to proceed or to proceed at all.

The purpose of testing, computer modeling, engineering analysis, probability studies, Failure Modes and Effects Analysis (FMEA), Fault Tree Analysis (FTA) and good old-fashioned thinking is to generate and evaluate information so that decisions can be made.

> *"However, we do not have the luxury of collecting informa-*
> *tion indefinitely, At some point, before we can have every*
> *possible fact in hand, we have to decide. The key is not to*
> *make quick decisions, but to make timely decisions. I have a*
> *timing formula, P = 40 to 70, in which P stands for probabil-*
> *ity of success and the numbers indicate the percentage of in-*
> *formation acquired. I don't act if I have only enough informa-*
> *tion to give me less than a 40 percent chance of being right.*
> *And I don't wait until I have enough facts to be 100 percent*
> *sure of being right, because by then it is almost always too*
> *late. I go with my gut feeling when I have acquired informa-*
> *tion somewhere in the range of 40 to 70 percent."* [1]

Colin Powell, in his autobiography, outlined his criteria for making decisions. He observed that in most cases (especially in his career in the military), the individual did not have the luxury of collecting 100% of the information needed to make a bulletproof decision. On the other hand, making a decision without sound information would be foolish. The question, then, was how to balance the gathering and analyzing of information against the timeliness of the decision being made.

Business and engineering decisions work the same way. Any business plan requires information in order to make sound decisions: marketing analysis to determine the number of high-speed routers that the market will bear; cost of production based on volume; cost of overhead; neces-sary retail price to make profit. The question: Should the high-speed router be mass-produced or built on a per order basis? Making a wrong decision can cost a company dearly; making the right decision can drive a company to profitability. Currently, as this chapter is being written, technology stocks are still down and flat after 18 months. The technol-ogy "bubble" burst because many companies and investors made deci-sions based on little or no pertinent information.

Early in my career, I conducted a cable pull test for a client. The infor-mation from the test was used to make a decision about whether the client would bid a job to supply steel cable that would meet certain

---

[1] Colin Powell, *My American Journey*, (Ballantine Books, 1995), pp. 380–381.

strength and elongation criteria. The test was conducted, and the cable met the strength requirement, but miserably failed the elongation criteria. The client was informed and promptly turned down the supply contract.

The next day, an error was found in the extensometer setup (the device that measures the elongation of the cable under load) and the true elongation of the cable was calculated. The client was called with the good news (24 hours late) that the cable did indeed pass. Since I had made the error, I got the dubious honor of calling the client and taking care of the corrective action on the error. When I called the client, I expressed the hope that the 24-hour delay in the correct information had not caused a problem. Of course, it had. The contract was awarded to a different supplier. This was an unfortunate but valuable lesson: the test results were not enough; the information had to come at the right time.

Consider the Challenger disaster:

1. The Commission concluded that there was a serious flaw in the decision-making process leading up to the launch of flight 51-L. A well-structured and managed system emphasizing safety would have flagged the rising doubts about the Solid Rocket Booster joint seal. Had these matters been clearly stated and emphasized in the flight readiness process in terms reflecting the views of most of the Thiokol engineers and at least some of the Marshall engineers, it seems likely that the launch of 51-L might not have occurred when it did.[2]

The report concluded that "....views of most of the Thiokol engineers and at least some of the Marshall engineers..." were ignored at some level. A management decision was made, but based on what facts, on what information? Because the information was ignored, the fact that out of round issues in the seating and o-ring seal in the solid rocket booster would cause the seal to fail at low temperature was clearly, and dramatically, forced into the consciousness of all those involved.

---

[2] Report of the Presidential Commission on the Space Shuttle Challenger Accident (in compliance with Executive Order 12546 of February 3, 1986).

The purpose of testing, computer modeling or any other information generator is to provide information and analysis so that a sound decision can be made. When the information does not match the decision, or the information is not available in a timely fashion, bad decisions are made.

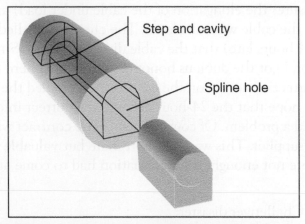

Figure 1-1: D-spline with spline hole bottoming out in a step and cavity for molding purposes.

This is true in business and in engineering. Over the years, I have done a wide range of engineering and testing. In one case, I was working on developing Entela's Finite Element Analysis (FEA) capability. We identified a job in which a component we were testing was consistently failing. We offered to conduct the finite element analysis and failure analysis in order to help identify the source of the problem. The design had already gone through several revisions and continued to fail. When we conducted the FEA, it was determined that the highest stress concentration was on the *inside* at the base of a "D" spline connection between a motor and a baffle. The previous designs had all focused on increasing the lip and rim thickness of the "D" spline. But that was not the source of the failure. By not having the correct information, the design team could only guess at potential solutions. By identifying the key piece of information needed to solve the problem, the design became simple to correct.

This illustrates a fact that is extremely important for all who interact with validation and testing information to be conscious of: it's not the test that is important, but the information.

"We have lots of data, but very little information."

—Julius Wang, DaimlerChrysler Corporation, July 9, 2002.

A perfectly executed test, with highly accurate results, does not help solve a problem or make the foundation for a good decision, unless it produces the correct information. In the case of the "D" spline failure, the mechanical load that the rim of the spline could handle was not the issue. The key piece of information was *where* the failure originated. It should be noted that if we had not offered to provide a different service to the client, the client would have continued to make a design change and test the part. Unfortunately, the test being conducted was a life durability test designed to demonstrate whether the part could survive a life. Knowing the part failed to meet a life requirement did not provide the key piece of information needed to fix the problem.

Again, *it's not the test that is important, but the information.*

I restate that fact for this simple reason: As a society (and I am thinking globally), we have come to equate "testing" with being "good." How many health and beauty aids make the claim "clinically tested?" A search on the internet returned more the 124,000 hits for that phrase. But an examination of what is meant by the claim quickly shows that it is a cliché. "Clinically tested" does not mean that anything has been proven. Think about that statement rationally for a minute. Just because it is "tested" does not mean that it is "good."

Go back to your high school science class. An experiment establishes whether a hypothesis can be disproved or not. An experiment never establishes, and no scientist or engineer or experimenter who really understands science will ever say, that an experiment proves that a hypothesis is true. Reject the hypothesis or accept it, never prove it true.

In business and engineering, we tend to equate the conduct of the test as certification that the product is good. I have seen countless project timelines in which the final validation testing was going to be conducted right up to the time when production would start. The implicit assumption was that the product would pass.

The book, *The McKinsey Mind*, by Ethan Rasiel and Paul Friga, details the structured thought process of McKinsey & Company, a top business consulting firm. The very first fact that they establish in chapter one is the need for a FACT-based analysis derived from a structured HYPOTHESIS-driven thought process.[3] They also note that many "McKinsey-ites" who leave the firm discover that many American firms have very poorly structured decision-making processes. The reality is that a test must be conducted on the basis of a hypothesis, and the hypothesis must be linked to a business or engineering decision.

If you can't state the hypothesis of a test, then it probably is not a test.

I asked a client who was working with Entela's engineers doing extensive testing on audio connectors what his timing requirements were. He said that they would go into production within the month. I asked what the plan was if the connectors failed. He said that they would go into production within the month. I asked what they would do different if the connectors failed: *nothing*.

Before you laugh too hard at such foolishness, remember, you are afflicted with the same blindness. We must test before we commit to hard tooling, before we go into production. Do you see the blindness in that statement? "We must test before we commit to hard tooling, before we go into production." The real statement should be, "If we have data to support the hypothesis that our business model is based on, then we will commit to hard tooling, and go into production. If not, then we will reformulate the business plan."

It's not the test that is important, but the information.

---

[3] Ethan Rasiel and Paul Friga, *The McKinsey Mind*, (McGraw-Hill, 2002).

If this blindness to the importance of the information is not real, then why does every project timeline I have ever seen for bringing a product to market include the time for testing, instead of the time *and* decision branch, for collecting and reacting to key information? The testing is supposed to be a tool, not an end unto itself.

"Time heals all wounds."

Time may heal all wounds, but entrenched misconceptions such as: "I tested it, therefore it's good" do not get better with time. They may change, morphing with the trends and subtleties of a complex society, but they do not get better without considerable effort on the part of a broad range of individuals. Take a look at how opinions in the testing community have changed over time. Read the preface from reliability and testing books circa 1990. You will find very confident statements such as:

> *"Reliability is the best quantitative measure of the integrity of a part, component, product, or system.*
>
> *Reliability Engineering provides the theoretical and practical tools whereby the probability and capability of parts, components, equipment, products, subsystems, and systems to perform their required functions without failure for desired periods in specified environments, that is their desired optimized reliability, can be specified, predicted, designed in, tested, demonstrated, packaged, transported, stored, installed, and started up, and their performance monitored and fed back to all concerned organizations, and any needed corrective action(s) taken; the results of these actions being followed through to see if the units' reliability has improved; and similarly for their desired and optimized maintainability, availability, safety and quality levels at desired confidence levels and at competitive prices."*[4]

---

[4] Deimitri Kececioglu, *Reliability Engineering Handbook, Volume 1*, (PTR Prentice Hall, 1991), p. 2.

Read the statement for its structure as well as what it says. There is a similar structure to more famous sayings throughout history:

> *"War to end all wars"* and,
>
> *"Everything that can be invented, has been invented."*
> —Charles H. Duell, Commissioner,
> U.S. Office of Patents, 1899.

Kececioglu is reflecting the prevailing attitude at the time—statistical quantification of performance is the best way to do everything. The fact of the matter is that statistics is only *one branch* of mathematics, and mathematics is only *one form* of communication. If the real goal is the correct information to make a sound engineering or business decision, then the tools (statistics, mathematics, failure analysis, physics of failure, fault tree analysis, DFMEA, FEA, design maturity) are all valuable, and different tools will be best at different times.

> *"Engineers are like kids, give a kid a hammer and everything needs pounding, give an engineer a new tool and it will be applied to everything."*[5]

As we move closer to the turn of the millennium, the prevailing opinion changes.

> *"Accelerated testing can be divided into two groups: qualitative testing and quantitative life testing. In qualitative, the engineer is mostly interested in identifying failures and failure modes without attempting to make any predictions. In quantitative life testing, the engineer is interested in predicting the life of the product at some normal use condition."*[6]

Here we see a decided change in opinion. No longer is statistics (quantitative life) the only means of gaining and relaying information. The blindness was morphing, and probably in a good direction, but be careful. Conducting a qualitative or quantitative test does not mean you

---

[5] "Accelerated Testing Seminar," by Alex Porter, Entela, Inc., 1999.

[6] "SAE Advances in Automotive Systems Testing Toptec," by Pantelis Vassiliou, ReliaSoft, May 7–8, 2002.

have collected good information; for example, there is lots of data, but is it information that is *needed*?

Let me offer a working definition of information for the purposes of this book: information is data that has been distilled into a pattern within a context that affects the behavior of sentient beings.

Data that informs a decision is information, data that doesn't, isn't.

All those memo's marked FYI are data, the call from your child's elementary school about a broken arm is information.

In addition to the change in perceptions about the "best" methods, the perceptions also change with the type of business. Entrepreneurs will often test only key points of a new, innovative design that they are unsure about. The value that they bring to the marketplace is the innovation, so demonstrating the performance of the innovation is often the focus of the testing. With an established commodity with lots of competition for essentially the same product, testing focuses on the cost of quality, reliability and efficiency of production. These are two very different information-generation needs based on the business model. In one case, testing is desired to highlight the unique new features of a new product (which is the focus of the business model); in the other testing, it is used to provide minute adjustments to design and production to improve reliability and price point (which is the key to success in the commodity, mass production business model).

Table 1-1.

Entrepreneur Bottom Line			
Fixed Overhead		$	150,000.00
Production Cost/unit	$ 50.00		
Production Volume	1,000		
		$	50,000.00
Destribution Cost	$ 2.00	$	2,000.00
Warranty Cost	$ 10.00	$	10,000.00
Sub Total		$	212,000.00
Sale Price	$ 550.00	$	550,000.00
Net		$	338,000.00

Consider the simple bottom line model for production of an innovative product. There is little competition, so the sale price has a large margin. It can be easily shown that the key factor for the margin and the business model is the degree of innovation that allows the large margin. A substantial change in cost of production or in the warranty costs does not have a significant impact on the bottom line.

Table 1-2.

Commodity Bottom Line		
	Unit	**Cost**
Fixed Overhead		$ 452,000.00
Production Cost/unit	$ 1.40	
Production Volume	1,000,000	
		$ 1,400,000.00
Destribution Cost	$ 0.30	$ 300,000.00
Warranty Cost	$ 0.10	$ 100,000.00
Sub Total		$ 2,252,000.00
Sale Price	$ 1.90	$ 1,900,000.00
Net		($352,000)

On the other hand, with a commodity product with lots of competition, the value is not the innovation but the price point. The margin is small, volumes must be large, and the effect of production cost or warranty cost per unit on the bottom line is very large. In the two simple examples shown, the production cost effect on the net is 10:1 for the innovative product (meaning a 10% change in production cost produces a 1% change in the net), while the commodity has a 1:3 ratio (10% change in the production cost results in approximately a 30% change in the net).

Naturally, these two types of products result in two different focuses for validation. With the innovative entrepreneur, the focus is on demonstration of the innovation, while the commodity must find small price point changes in production costs in order to realize a net profit.

The white goods industry is a good example of a commodity where a clothes dryer that sells for $300 has less than a dollar in margin. However, the white goods industry produces huge volumes and is extremely price point conscious. Some of the most interesting projects I have worked on were for consumer white goods testing projects.

On the other hand, certain high-end telecommunications or power management devices are very low volume, highly innovative. The cost of over-designing the cost of production when 1000 units will be produced is much smaller than the testing and validation that is necessary to ensure that a cost reduction does not change the durability of the product.

Consider this example: For a high volume production (10,000,000 units per year), a reduction in sheet metal gauge of one gauge size could result in 0.1 lbs. per unit reduction in raw material. Material cost of $0.50/lbs. will result in a savings of $.05/unit. That amounts to a $500,000/year savings applied directly to the bottom line. For a product with 1,000 units per year, this would be a $50 savings. What testing would be needed to ensure that the reduction in gauge size did not result in an increase in warranty cost (that both gauges would have the same reliability)? A life/durability study comparing the two gauges would provide the key information needed to make this decision. If the cost of this type of testing was $50,000 in time/material, then for the high volume production this information is useful; for the low volume production, it's meaningless.

Another factor that impacts the information needs of the decision-maker is the type of supply chain and the company's position in the supply chain.

I worked with one manufacturer that was a fully integrated manufacturing and distribution company. They designed, manufactured, marketed and serviced everything in their product. They even wound their own armatures in their motors. The reason for this business model was the need to control quality to a very high level to ensure a good reputation in their direct marketing sales approach. The decisions made about design changes, durability, reliability, and cost of quality were fully integrated and made by a team.

Compare this approach to the automotive supply chain where the U.S. OEM's are assemblers who purchase entire sub-systems from major tier one suppliers, who purchase components from tier two suppliers. The OEM is only interested in the top-level view and continues to push warranty, design and validation responsibility down to the suppliers. For the OEM, the decision is based on which supplier to choose and how the major systems interact (full vehicle). For the tier one supplier, the decisions are made about which tier two suppliers to use and the system level (component interaction). For the tier two suppliers, the decisions are about minute design details on individual products, their performance and durability. A test method designed for the OEM to ensure full vehicle Electro-Magnetic Compatibility (EMC) will be very different than a test method for a tier two supplier of a radio. The radio supplier may need the results of a very detailed functional and durability test in order to ensure that the radio works properly, but the tier one supplier (the system integrator) will only care about the radio bracket, heat dissipation, wiring interface and other integration issues.

One interesting human interaction that I have witnessed while working with companies on test plans, is the conflict that arises because of the various parties' information needs. The situation usually develops when a meeting is called to review a test plan designed by whoever holds the purse strings. The plan is presented to the team working on the project. Inevitably, somebody will ask if a certain measurement will be made, or a certain quantity will be determined. When the answer is to the negative, the conflict arises. For example, a reliability engineer commissions a test plan to determine the Mean Time Between Failure (MTBF) of an assembly. The plan is presented to a team that includes the reliability engineer, a design engineer, the warranty engineer and the production engineer. The design engineer asks if the optimal resistance for a key resistor in the power circuit can be determined: *no*. The production engineer asks if the sensitivity to dimensional variations of key dimensions of the plastic enclosure will be determined: *no*. The warranty engineer asks if the key warranty issues and their relative severity will be determined: *no*. Well, then what good is this test? I have been in meetings where this conflict is played out with different combinations of people, expectations and even companies in the supply chain.

We see then several interacting forces on the perceptions and application of testing and validation. Time morphs the paradigms and opinions of what is generally viewed as "best", while company structure, style, place in the supply chain and the stage of product development define what is needed. To truly have a handle on the interaction of testing and validation practice at a particular company, you have to look at its business structure (both internally and in the supply chain), the project stage and where it is in the flow of time.

Julius Wang and Richard Rudy offered the following pyramid of accelerated testing adoption at a 2002 SAE Toptec.

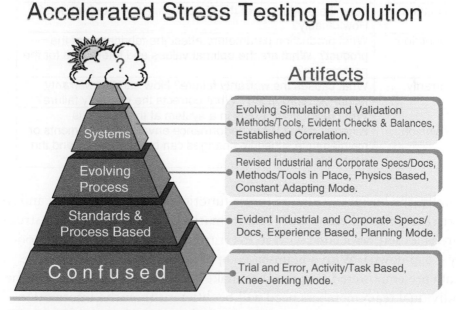

Accelerated Stress Testing Evolution

Artifacts

Evolving Simulation and Validation Methods/Tools, Evident Checks & Balances, Established Correlation.

Revised Industrial and Corporate Specs/Docs, Methods/Tools in Place, Physics Based, Constant Adapting Mode.

Evident Industrial and Corporate Specs/ Docs, Experience Based, Planning Mode.

Trial and Error, Activity/Task Based, Knee-Jerking Mode.

Systems

Evolving Process

Standards & Process Based

Confused

Figure 1-2: Accelerated stress testing evolution.

This evolution of accelerated testing applies equally well to the stage of a company. At the entrepreneurial stage, testing is conducted on an as-needed basis. As a product becomes established and begins more regular production, some standardization of test and process takes place. As the product becomes a commodity, the process evolves and is refined to improve reliability and price point. Some companies will rise to a true level of excellence and set the standard for the testing and process methods.

The pinnacle is purposely left undefined. Wang and Rudy assert that the ultimate in testing has not been reached as yet.

This progression in the application of testing tools based on the progression of the business model is also influenced by the development stage of a product.

Table 1-3.

Research:	What are the boundaries of a new type of technology?
Development:	What design features need correcting? What must be changed to make it work?
Validation:	Does the product meet the life/performance requirements? How reliably?
Production:	What production parameters affect the fabrication of the product? What are the optimal values and tolerances for the parameters?
Warranty:	What causes the warranty failure? How can the warranty failure be reproduced? What corrects the warranty failure?
Life Extension:	What residual life exists in a system at the end of its scheduled life? What performance envelope adjustments or maintenance schedule changes can be made to extend the useful life safely?

Finally, the business structure (cross-functional teams, top-down and so forth) requires different types of information. Often, the business structure is loosely associated with style (entrepreneurial – team; commodity – top-down). With a team-based approach where members of the team are empowered (required) to make decisions, the information the individual team members need will be different than the information needed by a top-level executive in a top-down business structure.

A testing scheme produces data. The data is interpreted in the context of a particular situation based on:

Table 1-4.

Management Structure:	What information is needed for different levels in the structure to make decisions?
Business Style:	What is the business model and what information drives it forward?
Place in Supply Chain:	What is the supplier level (OEM, Tier One, Tier Two)?
Product Phase:	What information is needed at each stage of product development, production and life to make good decisions?
Time/Money/Risk:	What testing will provide the information needed within the time and resource constraints to minimize the risk to business decisions?

Accelerated testing comes about because traditional testing methods often fail to meet the needs listed in Table 1-4. However, the newest accelerated test does not mean that it is the best or most appropriate for a particular situation. Forming a clear understanding of the information needs (data, context and time) and keeping that as the foremost requirement can help keep an organization conducting tests for their information and not simply to say, "We tested the product."

The reality is that data exists all around us. Most likely you have a hard drive in a computer (or on a shelf) near you. There is data on the hard drive, some of that data might be useful. But the data can not be comprehended directly by a human being. We have no biological mechanism for consciously detecting magnetic fields. The information on the hard drive must be converted to a form that can be sensed by one of our senses (sight, sound, smell, touch, taste) before we can comprehend the information.

Data exists in many forms all around, and much of the data would be useful information for making engineering and business decisions if it could be precipitated out of its environment into a form that we could sense. With a hard drive we use a magnetic read/write head, software, displays, speakers, printers, and Braille to convert the data to information that can be sensed.

Consider the case of an automotive stalk switch (that knobby stick on the steering column that has the wiper, turn signal, and cruise control on it). In one particular case, the wiper motor speed was communicated to the automobile's computer through an impedance level on the switch. There were certain conditions under which the wiper motor would spontaneously run one intermittent wipe and then stop. This was dubbed the "phantom wipe" and was causing some minor annoyance for the automobile owners (and the dealers trying to address the warranty claim).

The reality was that the *data* existed right in front of them in the form of the product design and its reaction to the environment. But that information could not be comprehended until the product was subject to particular controlled environments (and in this case, in a particular sequence) to produce the condition. The information (data with context that influences sentient behavior) had to be precipitated out of the data (material composition, part design, environmental conditions, software) so that it could be recognized and interpreted. The information existed all along, but not in a form that could be recognized.

Before Mt. Everest was discovered, Mt. Everest was the tallest mountain; the information existed, but had not been precipitated through exploration and measurement into a form that could be cognitively recognized.

Cognitive recognition of patterns in data within a context that influences our behavior is the act of *recognizing information*. However, the fact that our minds are involved and the patterns that we see in the data are a part of forming data into information means there is some inherent bias in any data interpretation. Think about why medical studies use placebos as controls. Our minds insist on putting data in order, even when order may not exist.

I was consulting for a company in Canada that was making Exhaust Gas Recirculation (EGR) valves. This is one of those expensive little gadgets that helps keep cars environmentally friendly. The valve receives a signal from the computer telling it how far to open, and then sends a signal telling the computer how far the valve is open. Testing the performance of the valve is a simple matter of sending a signal to the valve to adjust its position and then monitoring the position signal the valve sends while under a variety of environmental conditions.

When we were setting up to run a new test on the valve, we put an oscilloscope on the input and output of the valve (precipitated voltage into a visual form). The scope showed the square wave that was being sent into the valve. The output of the valve showed a less than perfect response.

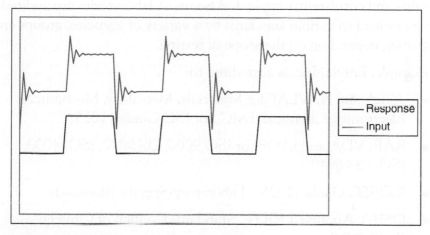

Figure 1-3: Input and response of EGR valve.

When I asked the engineers what the noise was on the valve response, they replied that it was nothing—just some noise. Square wave in, square wave out was their paradigm and they ignored the data that did not fit. I went out on a limb and said that the part will fail in a way that was associated with the noise at the rise and fall of the square wave. Fortunately, I was correct; as the parts went through various environ-ments this noise got worse and worse until the valve failed. The reason I was willing to go out on the limb was simple: It was obvious that they were ignoring key data in order to get the information they wanted— square wave in, square wave out. The failure, then, was likely associated with the data that was precipitated into information, but ignored.

So what do we do with our testing schemes in order to avoid this effect (called the *"paradigm effect"* from the video, *Paradigm Principles*, by Joel Barker) of forcing data and ignoring data to get the information we want?

In recent years, quality systems (ISO-9000 being a well-known example) have attempted to provide objective controls on a wide variety of aspects of commerce, business, engineering, manufacturing and so forth. For test laboratories, ISO Guide 25, and more recently, ISO Guide 17025 outline the process controls and procedures that should be in place to ensure that testing practices are validated, documented, traceable and consistently applied. Also, most laboratories are audited and/or certified to various standards by a variety of agencies, groups and companies, depending on the scope of testing.

For example, Entela, Inc. is accredited to:

- A2LA and NAVLAP for Materials, Metrology, Mechanical, Electronics, Chemical, EMC, for ISO Guide 17025

- RAB, VDA and IAOB for ISO 9000, QS 9000, ISO14000, ISO/TS 16949

- ISO/IEC Guide 17025 – Laboratory-Specific Standards

- OSHA-Approved NRTL, Standards Council of Canada, IEC/CB Scheme

While quality systems and auditing can help reduce the effect of bias on interpreting test data, the personal discipline of the producers and users of testing data is the real defense.

Historical Business Models and the Information Needed

Historically, businesses have been formed around the top-down business structure borrowed from military command structures. The basic premise of the top-down business structure was that there is a person responsible for everything and that responsibility and authority would be delegated down from the top.

This structure means that middle- and lower-level employees in the structure had little or no authority—they were not supposed to think or make decisions, just perform. Because individuals at the upper levels of the company are making the decisions in a top-down business structure, the decisions tended to be about aggregate, top-level issues. How many warranty parts should be inventoried? How long should the warranty period be granted? The top level of management did not worry about details (what radius fillet should be placed on this corner?).

Because the top level of a top-down business structure was making the decisions (and controlling the purse), the information generators were serving their needs. Testing and validation provided aggregate information on the population behavior of a product (reliability, MTBF, cost of quality). These parameters also fit with the prevailing notion that statistical quantification was the only way to measure.

No cognitive recognition at lower levels (check the box).

This meant that lower levels were required to simply do what they were told. Produce so many parts, measure the reliability with a certain confidence, ship the parts. It was not required of them to think about why or how.

For the top-down structure, time is controlled, flow-charted, Gant charted and so forth. The information generated is needed by the top level at precise points in time in order for good decisions to be made.

Working Group Structure (Entrepreneur)

Figure 1-4: Project team meeting on MEOST testing for Whirlpool Corporation brought engineers from several plants across the country together to brainstorm information goals and test plans.

More recently, companies have used working groups or cross-functional teams. In reality, this is not new. Most innovative products start out being developed in cross-functional teams and the resulting company migrates to a top-down structure after establishing the product as a commodity. In a cross-functional team, all lines of communication are open and all levels are encouraged to think and make decisions. For practical purposes, there is still someone responsible at the top, but their role is shifted from dictator to facilitator. Their job is to facilitate the flow of communication between the members of the team and to keep the team focused on the goals.

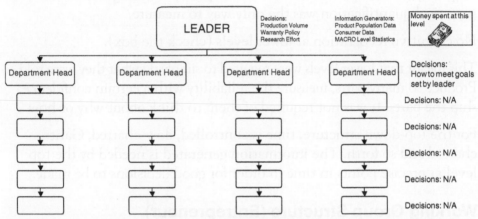

Figure 1-5: Top-down structure. The decisions are made at the top, information generators provide information for those decisions.

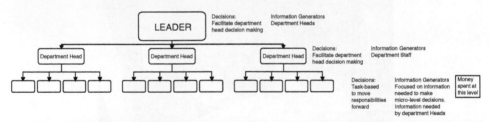

Figure 1-6: Cross-functional teams. Decisions are made throughout the structure, information generators provide information needed.

The information needs for the cross-functional team may be much different than that of the top-down structure. Take a team developing a high-speed digital router. Digital, power, software and enclosures must all work together. While a top-down manager would want information generated on performance margins and the product's population reliability, the cross-functional team may need to know the failure mechanisms of the power circuit, the fault tolerance of the solid state electronics or the durability and heat dissipation of the enclosures. These are detail-oriented information needs instead of aggregate result-oriented information. The reason is simple: the cross-functional team must make decisions about designing and producing the product, while a top level top-down manager must make decisions about cost, warranty terms, and managing the product population.

Telling the cross-functional team that the current design has an MTBF of 48 months would be data, not information. How do you change a power circuit based on a statistical measure of the population's aggregate performance? The population's performance is influenced by a wide variety of factors, the power circuit being only one. Then again, how does a top-level manager plan warranty terms based on a power circuit having its weakest member in capacitor C4? The cross-functional team could make decisions about the product based on detailed information like an incorrectly sized capacitor in a bridge, while a top-level manager can make planning decisions based on statistical performance of a population of products.

Modern Business Models and the Information Needed

Here is an interesting challenge: Suppose you have a large corporation, one of the largest in the world. For years it has been managed in a top-down business structure. The information generators (testing, computer modeling and so forth) and their well-documented specifications and procedures (which are controlled by the quality system) support the generation of information needed to make decisions in the top-down business structure. Now change to cross-functional teams in the development and manufacturing process of the company structure. The members of the team are responsible for making decisions....but wait, were the information generators (still controlled by the quality system) changed to reflect the change in information needs?

One good thing about a quality system is that important procedures were documented and consistently followed. However, changing a thoroughly documented and indoctrinated practice in a large company is very difficult. Now you have information being generated that the top levels of the company recognize and have used for years to make decisions, but there are cross-functional teams at lower levels that are now responsible for making aspects of the business model work for whom the data generated does not produce much useful information.

So when does the team get the information they need? When the lack of information produces serious problems (warranty), and upper-management releases funds to "do what ever it takes" to get the situation under control. The team is then free to employ information generating tools to get the key pieces of information needed.

Key facts going forward:

1. Information is data, interpreted in a context that influences the behavior of sentient beings.

2. It is not the test that is important, but the information.

3. The information only has value if it is available at the correct time.

4. Eighty percent of the information in time to make a decision is far more valuable than one-hundred percent of the information 24 hours late.

5. Good information for one situation is just data for another.

6. Different business models, styles and supply chain positions require different information to make good decisions.

7. Different levels of development require different information to make good decisions.

8. A test designed to quantify one characteristic will not necessarily quantify a different characteristic. (Remember Werner Heisenberg?)

9. Perceptions of what is "best" at present will change as the business models, styles, and stages of development change. Remember point two above.

Precise But Useless Data

This chapter deservers a bad joke...

> *A balloonist arranges for transportation for himself and his*
> *balloon at a precise place. After flying for some time, he*
> *realizes he's lost. He reduces height and calls to a man on the*
> *ground, "Excuse me, I was supposed to meet a friend half an*
> *hour ago, but I'm not sure of my location?"*
>
> *The man on the ground says, "You are in a hot air balloon,*
> *hovering approximately 30 feet above this field. You are be-*
> *tween 40 and 42 degrees north latitude, and between 58 and*
> *60 degrees west longitude."*
>
> *"You must be an engineer," the balloonist says in frustration.*
>
> *"Yes, I am," replies the man. "How did you know?"*
>
> *"Everything you have told me is correct, but you haven't re-*
> *ally helped me, in fact, I am still lost."*
>
> *The man on the ground says, "You must be a manager."*
>
> *"Yes, I am," replies the balloonist. "How did you know?"*
>
> *"You don't know where you are, where you're going or how*
> *to keep your promises, and after one question it's all my*
> *fault."*

The engineer in this old joke gave very precise data to the manager
based on traditional methods of measurement. After all, latitude and
longitude have been used since the 17th century, and feet as a measure

of length have been used since early England. The results of the engineer's estimates will undoubtedly bear up to third-party verification and any audit (2 degrees range covers a large area). The data has context (see point one from Chapter 1), but it does not provide any means of influencing behavior. It is just data, not information.

Many traditional validation tools can run into the same problem of application as the location data provided by the engineer. Very precise and repeatable, but completely useless.

Accurate But Not Beneficial

I recall a phone call, actually several phone calls, from engineers, technicians or purchasing agents asking for "testing".

Caller: "I have a widget that I need tested."

Test Engineer: "Is there a specific standard or requirement you would like to conduct the testing to?"

Caller: "I don't know. Whatever is the normal test."

Test Engineer: "For a widget?"

Caller: "Yeah."

Test Engineer: "It depends on why you are testing the part."

Caller: "Because I have to."

Long silence as the Test Engineer takes a deep breath and counts to 10.

Test Engineer: "What will you do with the information?"

Caller: "File it."

A quick survey of the usual suspects for "standard" test methods provides a short list:

- SAE (Society of Automotive Engineers) – Automotive, Aerospace and Transportation

- IEEE (Institute of Electrical and Electronics Engineers)

- ASTM (American Society for Testing and Materials)

- Telcordia Technologies (Bellcore)
- ISO (International Organization for Standardization)
- IEC (International Electrotechnical Commission)

In addition, there are government and company specific methods:

- DOT (Department of Transportation)
- DOD (Department of Defense)
- FDA (Food and Drug Administration)
- GMW (General Motors Worldwide)
- FMC (Ford Motor Company)
- Boeing
- CNS (Chinese National Standards)[1]
- JISC (Japanese Industrial Standards Committee)
- DIN (German Standards)

Many of the standard tests from one organization will reflect the corresponding standard test from another organization. For example, the DOT automotive lighting standard 108 is closely based on the SAE headlamp specifications. In industry, the individuals who work on one committee may coordinate with or participate on another committee. This provides some continuity of methods between organizations; however, it does limit innovation and dramatically increases the momentum for one particular method.

All of the different specifications cover a range of goals. For example, some specifications are designed to standardize common mechanical or electrical elements or practices. Standards for "hook and loop" fasteners (commonly known as Velcro® [2]) will detail the minimum requirements a product must meet to be called an "ASTM compliant" hook and loop fastener. These types of standardizing specifications provide the basis for commonality in fasteners, material designations and so forth. Other

---

[1] Taiwan.
[2] Velcro Industries B.V.

standard specifications detail how certain pieces of information are to be collected. ASTM E8 provides the standardized methods for determining metal yield and tensile strength and elongation. Other standards detail the testing of functional reliability and life durability.

In examining the past and current validation methods, how they are used, how they are misused, and how they may be accelerated, it is important to have a clear map of the variety of types of specifications used as tools in the validation and testing process. Following are some examples:

- Functional/dimensional specification
- Functional feasibility
- Functional reliability
- Compatibility
- Life reliability
- Stresses/conditions
- Properties

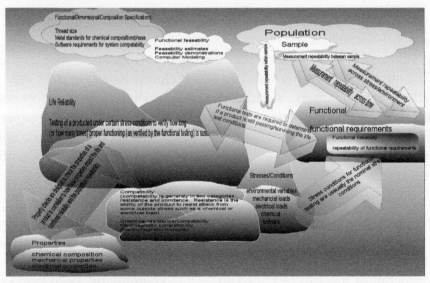

Figure 2-1: Testing type map.

Commonality specifications (functional/dimensional/composition) such as SAE 40 weight motor oil, ¼-20 coarse threaded fastener, or 316 stainless steel are specifications that determine the minimum characteristics for products that must be common from one manufacturer to the next. Often, these are tried and true dimensions or characteristics that have come down from past generations and have been coded into standards and sometimes law by materials, engineering or chemistry organizations and governments. The specifications detail what the common elements must be, and often how the characteristics are to be measured. The type of tests found in these specifications are usually not long in duration and are not the subject of much acceleration. They are useful, however, when accelerating entire validation plans (see Chapter 13).

Feasibility tests are tests designed to determine if something is possible. The feasibility test does not attempt to prove that a given design WILL work, or that it HAS worked, just that it is possible. This can be done as a paper study using assumptions ("given past success with removing metal from cross-car beams in automotive cockpit design and assuming X, Y and Z where accomplished…") to extrapolate ("an all plastic cross beam would be possible…") the possibilities ("to be constructed with a mass 20% less than current models.").

Computer modeling of current products compared to proposed designs provides a slightly more rigorous extrapolation. See Chapter 11 for a discussion on the inherent assumptions in computer models.

Life reliability testing or durability testing is testing of a product under certain stress conditions to verify how long (or how many times) proper functioning (as verified by the functional testing) is sustained. This test usually ends up being very long, very difficult and very expensive. Most accelerated testing will focus on this type of testing.

Life reliability tests require several other types of testing. These tests are used to determine and verify the characteristics of a product as it progresses through the life test. Properties such as chemical, mechanical or electrical may be measured before, during and/or after the life durability. Changes in these properties may be used to quantify degradation or failure of the product. Compatibility to certain stresses such as chemical attack or electromagnetic fields may also be used to evaluate a product before, during and/or after the life testing. Stress conditions or noise factors (sources of potential damage to the product) are imposed during a life test.

Functional testing during the life reliability test is critical. The functional test (along with the properties and compatibility) defines the criteria for what a "good" part is and what a "bad" part is. Functional testing is often done at the "nominal" stress conditions.

Along with functional testing comes different sampling techniques. These techniques measure the repeatability within samples (how well does a given sample at a given state in its life repeat the same functional measurement, i.e., 20 resistance measurements on a single resistor); repeatability between samples (how comparable are the functional measures on a group of samples at a given state in their life, i.e., resistance levels measured on 20 resistors); repeatability across time (how well does a given functional measure repeat over time during the life of a product, i.e., resistance measured on a resistor once a year for 20 years); repeatability across stress (how does a given functional measure change over different stress conditions, i.e., resistance measured on a resistor at 5 degree intervals from 0 °C to 100 °C).

To illustrate this, consider the following U.S. government specification for headlamps from **49 CFR Ch. V (10–1–01 Edition)**:

> **§ 571.108 Standard No. 108; Lamps, reflective devices, and associated equipment.**
> S1. *Scope.* This standard specifies requirements for original and replacement lamps, reflective devices, and associated equipment.

With the exception of feasibility, each type of testing described above is used in the U.S. government lighting standard.

Durability:

S8.8 *Vibration.* A vibration test shall be conducted in accordance with the procedures of SAE Standard J575e *Tests for Motor Vehicle Lighting Devices and Components* August 1970, and the following: The table on the adapter plate shall be of sufficient size to completely contain the test fixture base with no overhang. The vibration shall be applied in the vertical axis of the headlamp system as mounted on the vehicle. The filament shall not be energized.

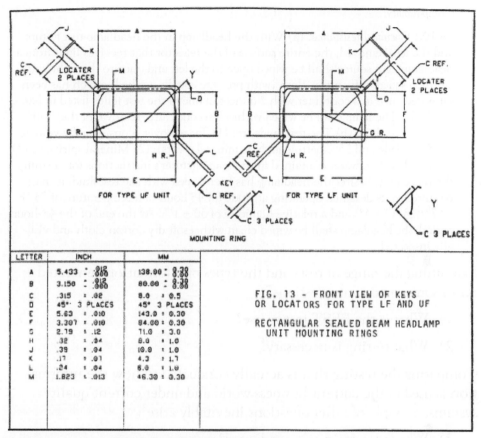

LETTER	INCH	MM
A	5.433 ± .012/.000	138.00 ± 0.30/0.00
B	3.150 ± .012/.000	80.00 ± 0.30/0.00
C	.315 ± .02	8.0 ± 0.5
D	45°· 3 PLACES	45° 3 PLACES
E	5.63 ± .010	143.0 ± 0.30
F	3.307 ± .010	84.00 ± 0.30
G	2.79 ± .12	71.0 ± 3.0
H	.32 ± .04	8.0 ± 1.0
J	.39 ± .04	10.0 ± 1.0
K	.17 ± .07	4.3 ± 1.7
L	.24 ± .04	6.0 ± 1.0
M	1.823 ± .013	46.30 ± 0.30

FIG. 13 - FRONT VIEW OF KEYS
OR LOCATORS FOR TYPE LF AND UF

RECTANGULAR SEALED BEAM HEADLAMP
UNIT MOUNTING RINGS

Figure 2-2: LF and DF type headlamp dimensional specification. Dimensional specifications do not convey quality but commonality within an industry.

Functional:

S8.9 *Sealing.* An unfixtured headlamp in its design mounting position shall be placed in water at a temperature of 176 ± 5 °F (60 ± 3 °C) for one hour. The headlamp shall be energized in its highest wattage mode, with the test voltage at 12.8 ± 0.1 V. during immersion. The lamp shall then be de-energized and immediately submerged in its design mounting position into water at 32 +5 –0 °F (0 +3 –0 °C). The water shall be in a pressurized vessel, and the pressure shall be increased to 10 psi (70 kPa), upon placing the lamp in the water. The lamp shall remain in the pressurized vessel for a period of thirty minutes. This entire procedure shall be repeated for four cycles. Then the lamp shall be inspected for any signs of water on its interior. During the high temperature portion of the cycles, the lamp shall be observed for signs of air escaping from its interior. If any water occurs on the interior or air escapes, the lamp is not a sealed lamp.

Compatibility:

S8.10.1 *Chemical resistance.* (a) With the headlamp in the headlamp test fixture and the lens removed, the entire surface of the reflector that receives light from a headlamp light source shall be wiped once to the left and once to the right with a 6-inch square soft cotton cloth (with pressure equally applied) which has been saturated once in a container with 2 ounces of one of the test fluids listed in paragraph (b). The lamp shall be wiped within 5 seconds after removal of the cloth from the test fluid. (b) The test fluids are: (1) Tar remover (consisting by volume of 45% xylene and 55% petroleum base mineral spirits); (2) Mineral spirits; or (3) Fluids other than water contained in the manufacturer's instructions for cleaning the reflector. (c) After the headlamp has been wiped with the test fluid, it shall be stored in its designed operating attitude for 48 hours at a temperature of 73 °F ±? 7°?(23 °C ± 4°) and a relative humidity of 30 ± 1%. At the end of the 48-hour period, the headlamp shall be wiped clean with a soft dry cotton cloth and visually inspected.

Examining the range of tests and the types of information generated raises some important questions:

1) What is the BEST test to use?

2) What testing is necessary?

Comparing the testing that is actually conducted to how the information is used in the current business world and under current quality systems, a couple of other questions inevitably arise[3]:

3) Why is a certain test conducted?

4) How does it fit into the business model or quality and reliability plan?

Unfortunately, these questions are usually asked and answered without a lot of serious thought. One reliability engineer from General Motors explained it this way: Validation engineers are entry level positions and when they start the job they ask the first two questions. Their boss (who was the entry level validation engineer from a few years before) pulls out examples of projects that they worked on with the tests, validation plans and procedures from a couple of years earlier (which they likely got from their boss). They instruct them to take the information, modify it slightly for the given application and run with it. The last

---

[3] Porter, A., "Quality Begins with Good Data," *Quality Magazine,* May 2001.

two questions are never asked by the entry level engineer and are rarely asked by higher level managers.

Occasionally, a company or a professional organization will form a "Best Practices" committee to determine what the best testing and validation procedures are. I have consulted for and participated on a couple of these committees and invariably they run into some significant problems:

1) The committee is charged with examining the test practice only. Unfortunately, the users of the information (managers, design engineers, manufacturing engineers) often need to change their practices in order for a change in the test practice to be affective. Because the committee's mandate is on the test practice and does not include how the information is used, they are forced to focus on producing exactly the same result in the same form as older test methods so that the consumers of the information recognize the data. When a company has changed its business model or structure (but the users of information are still used to seeing data produced for the old business model), it is nearly impossible to effect a change in the test practice.

2) There is no one BEST test practice. As we discussed in the first chapter, the business structure, the business model and the stage of product development all impact the information needs. What is BEST for one situation is not best for another. Most Best Practices committees are formed around expertise in testing— not in the business and management side; this tends to blind the committee to these facts.

3) Risk aversion. Statistical evaluation of certainty and uncertainty of a test result are only half the risk. The other half is the HISTORY that a company or an industry has with a particular test. A highly subjective test that is prone to error (such as the Izod, notch impact test ASTM D256) can be shown to be statistically inaccurate, but with a long history of use and results, perceived confidence in the test will create a level of confidence in the test, not because it is accurate, but because companies and institutions have a history of interpreting and understanding the

information. This can be called the *momentum of history*. When a new test comes along that is objectively a better test, there is a serious barrier to adoption because of the *perceived* lack of risk with the older test. Even though one execution of the newer test may have less risk when compared directly with one execution of the older test, the newer test is compared against the complete historical body of knowledge of the older test.

These barriers result in slow or little change in test practices without a huge effort. The result is that test methods continue to be used and trusted even though the results may not be relevant to the business model and the method may not be the most effective tool available.

The challenge with the range in type of testing and in volumes of test methods for each type is determining which test is the correct test and what set of tests provides a concise and complete set of results. The tendency is to choose a test based on expediency of choice and not efficiency of test.

Several factors lead to the selection of tests that provide accurate data that is completely useless. One reality is experienced by this author as I work on this chapter. The wealth of choices in test methods and types of testing threatens to overwhelm my ability to get a grasp on the full scope of test methods available. In fact, I know that I have purposely limited the scope of testing methods and techniques to make the task manageable. I have been in the testing industry for over 15 years, have tested everything from dental implants to torque converters for the space station. I have utilized every type of testing described in this chapter in five different industries. And I still struggle to comprehend the range and scope of all of the test methods. Many validation and testing engineers with only a few years of experience or just out of college simply do not have the scope of experience to choose the best method. Therefore, they follow the example of those who went before them, simply using the methods that were used before.

Another reason a particular method is used is because of economic forces. Prototype sample cost, contractual requirements and available time all serve to drive testing choices. For example, suppose a product is to be produced by company A for company B. The purchasing company (B) wants a reliable product and specifies a given reliability in the purchase agreement, say 95% reliable with a 70% confidence under a set of defined conditions for the warranty period.

- 95% reliable 70% confidence

- Temperatures between 80 °C and –40 °C, with ramp rates of 5 °C/minute

- Nominal voltage of 120 vac

- 500 hours is equivalent to 3-year warranty

Company A must demonstrate that this product meets this requirement. Several techniques could be used to accomplish this. A simple reliability test using a number of samples "n," and testing until all parts fail under the conditions would result in a probability distribution. The reliability can then be calculated with a given confidence based on the number of samples n tested.

However, company B also wants the production to start within a month of the start of testing. 500 hours of testing would take 21 days. But the simple reliability test would take the n samples to failure. A product that lasts 500 hours with a 95% reliability is going to have all or nearly all of the n samples still function at 500 hours. Depending on the distribution, it may take 5–6 times the reliable life to fail even half of the samples. (See Chapter 8.) This requires Company A to use a fully censored reliability test. N samples to one life with no failures.

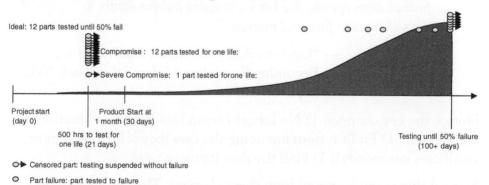

Ideal: 12 parts tested until 50% fail

Compromise : 12 parts tested for one life:

Severe Compromise: 1 part tested for one life:

Project start (day 0)

Product Start at 1 month (30 days)

500 hrs to test for one life (21 days)

Testing until 50% failure (100+ days)

○▶ Censored part: testing suspended without failure

○ Part failure: part tested to failure

Figure 2-3: The ideal statistical test and two compromises.

Now consider the usefulness of this test. It satisfies the contractual requirement of company A, demonstrating that the product is measurably 95% reliable with a given confidence under set conditions. The problem is that often company A will only do the contractual required tests instead of identifying the tests necessary for efficiently providing the information needed to successfully develop and produce the product. The contractual requirement is fulfilled, so the business side appears to be addressed, but sound engineering development has not taken place.

Add the economic pressure of reducing the sample size because of prototype costs and the statistical tests can become tragically flawed. One part tested for one life does not provide any statistical confidence but only demonstrates that it is possible to produce a part that can meet a narrow set of conditions for a fixed period of time.

So how real are the changes in business structures and economic forces? Consider this excerpt from *Information Week*:

> *"The power of Amazon.com and the lessons to be learned are in its structure and its reach across sectors. Believe it or not, the bastions of the Industrial Age—companies such as General Motors Corp. and Ford Motor Co.—compete as E-communities. Their structures are more Amazonlike than the hierarchies of old industries. These automakers earn much of their profit from financing rather than selling new cars. Their manufacturing units have been forced to almost halve development time to compete with Japanese carmakers. Their finance arms operate like banks, reacting independently to daily changes in financial markets."*
>
> — "Upstarts Alter the Rules, Businesses are Becoming Sectorless," *Information Week 500*, Sept. 11, 2000.

Notice the key changes: 1) No longer hierarchies but cross-functional and flexible; 2) Profit is from financing the cars they sell (in fact most small cars lost money); 3) Half the development time.

Several things can be noted from these changes. The automakers are highly conscious of price point, time to market is critical, development is done in cross-functional teams. One project I worked on for an automotive cockpit (from the front of the dash to the firewall, from the windshield to the floor) involved thirty different tier two suppliers, one tier one supplier and an OEM. The number of individuals and companies competing for validation results was astounding. Plus, the timing was such that the traditional testing that the OEM usually used would have had results after the second design freeze. You read that correctly: the development was to move so quickly that the design would go through two complete iterations in the time it would take to test the first prototype using the traditional tools. This would make the results from the testing absolutely worthless.

Automotive companies are going through relatively large changes, but they have a large amount of historical momentum in the test practices they are using. Newer companies have the luxury of choosing the best method they can find. I taught a seminar on accelerated testing in Seattle that an engineer from Microsoft attended. He was working on hardware (keyboards, mice and so forth). We discussed his current testing practices. Not surprisingly, they were modeled after the typical entrepreneurial software test plans: investigate the functionality and that's about it. Implementing modern test methods in this situation is relatively easy—there are no preconceived ideas of what information should be generated or how it will be used. Compare that to the thirty suppliers from the preceding cockpit example and you can see the magnitude of the difference.

One other historical reason for older test methods' longevity is reflected in a conversation I had with Milena Krasich, who is heading an exploratory committee for an accelerated testing standard for the International Electrotechnical Commission (IEC). Milena points out that the older style of specifications from the IEC were fixed format. The specifications stated what to do and what the results meant without any background or theory. The newer IEC specifications are including background, rationale, theory and examples so that people can determine how to apply the specifications and when the background assumptions are no longer applicable to the specific case. The newer specifications will also include limitations on the described method.

Having worked with specifications from ASTM, SAE, ISO and corporate sources, the older IEC style of being dictatorial is common. This style precludes any thinking by people in the trenches using the tools. Compare this to the top-down hierarchy of older industries and you may recognize the effects of the business model in the construction of the test methods. The top decides what to do, and once that is decided, the validation engineers simply carry out the tests and produce the results. This works fine for a top-down business model with years to develop product instead of months. However, in most cases the business model is long gone, but the effects of it are still entrenched in the volumes of standards and practices enshrined by businesses and institutions.

Precise Test

Consider a common form of durability or life test. Three samples to three lives with no failures. First, the particulars of the test plan, some theory and then the relevance to the business model and the decisions that needed to be made.

With the Device Under Test (DUT) mounted in the vehicle position:

1) Condition the part at 80 °C for 24 hours.

2) Apply the PSD from Table 2-1 for a period of 12 hours in each of three mutually perpendicular axis for a total of 36 hours. Repeat three times for a total of 108 hours per part.

3) During each 12-hour vibration cycle, apply the thermal profile from Table 2-2 and the electrical cycle profile from Table 2-3.

Table 2-1.

Hz	Grms^2/Hz
0	0
50	0.01
100	0.001
100	0.0005

Table 2-2.

Time (hrs)	Temperature (Deg C)
0	25
0.5	−40
2	−40
3	80
5	80
6	−40
8	−40
9	80
11	80
12	25

Table 2-3.

Time (seconds)	Power	Fan Speed	Vent Position
0	14	Low	Vent
20	14	Low	AC
40	14	Low	Defrost
60	14	Low	Defrost/Front
80	14	Low	Front/Floor
100	14	Medium	Vent
120	14	Medium	AC
140	14	Medium	Defrost
160	14	Medium	Defrost/Front
180	14	Medium	Front/Floor
200	14	High	Vent
220	14	High	AC
240	14	High	Defrost
260	14	High	Defrost/Front
280	14	High	Front/Floor
300	14	Off	Vent
320	14	Off	AC
340	14	Off	Defrost
360	14	Off	Defrost/Front
380	14	Off	Front/Floor
400	0	Off	Vent
Repeat 3 times per hour			

Each of the three parts must meet the following criteria while at room temperature and 14 volts of power before, during and after each 12-hour vibration cycle. No failures are allowed.

Table 2-4.

Vent Position	Blower Speed	Minimum Air Flow (cfm)
Vent	Low	10
AC	Low	10
Defrost	Low	10
Defrost/Front	Low	7
Front/Floor	Low	7
Vent	Medium	15
AC	Medium	15
Defrost	Medium	15
Defrost/Front	Medium	10
Front/Floor	Medium	10
Vent	High	20
AC	High	20
Defrost	High	20
Defrost/Front	High	15
Front/Floor	High	15

Traditionally, the preceding test plan would be presented with no explanation. But for this discussion, we will delve into the assumptions behind the test plan.

First, consider the vibration profile. This type of profile would be developed by collecting 3-axis (lateral, longitudinal and vertical directions) vibration data from a sample vehicle. Each vibration profile would look somewhat different. Out of a 15-year or 150,000 mile expected life, the "worst case" vibration would be identified for each axis. For example, a product might be expected to see 100,000 miles of PSD A, 30,000 miles of PSD B, 19,500 miles of PSD C, and 500 miles of PSD D. The worst condition is PSD D driven at 40–45 miles per hour. It is assumed that the worst case PSD represents the majority of the damage—so 12 hours at PSD D is assumed to represent one life. But, PSD D is a 3-axis vibration profile and most vibration equipment is single axis. Therefore, three 12-hour profiles will be run each in a mutually perpendicular axis. To simplify the test plan, the "worst case" PSD would be taken. This practice also provides a hedge against the assumptions made. The thinking is that a reasonable profile has been chosen and when compromises needed to be made, the compromise was always towards a harsher environment. This would be assumed to be the one with the greatest stress level. See Chapter 10 for an explanation of how lower energy can sometimes be worse on a part.

Now that a vibration profile has been determined, a thermal profile is estimated. Three criteria are used: maximum temperature, minimum temperature and ramp rate. Also, the electromechanical profile is determined. Again, some assumptions are made. How many cycles in one life? If you assume the average user adjusts the settings once a trip with an average of 430 adjustment changes a year, that is, 6,450 adjustments in one lifetime. This gives 179.2 adjustments per hour for a 12-hour life test, or 2.987 adjustments per minute or one every 20 seconds.

Finally, why three parts to three lives? This is a compromise based on some assumptions about distributions, strength, resources and time. Three parts to three lives takes longer to run (three times longer) but requires fewer prototypes. The statistical premise is that the confidence that zero failures in twelve parts gives if run for one life can be achieved by running only three parts for three lives given some assumptions about the shape of the population distribution. In other words, it is assumed that the likelihood of twelve parts lasting one life without failure is the same as three parts lasting three times as long. This is true for only one population shape.

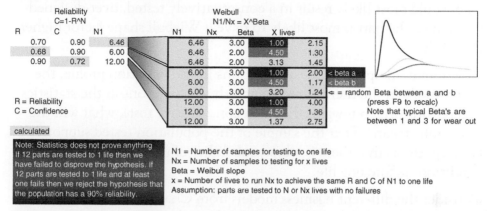

Reliability C=1-R^N			Weibull N1/Nx = X^Beta				
R	C	N1	N1	Nx	Beta	X lives	
0.70	0.90	6.46	6.46	3.00	1.00	2.15	
0.68	0.90	6.00	6.46	2.00	4.50	1.30	
0.90	0.72	12.00	6.46	2.00	3.13	1.45	
			6.00	3.00	1.00	2.00	< beta a
			6.00	3.00	4.50	1.17	< beta b
			6.00	3.00	3.20	1.24	<= = random Beta between a and b
R = Reliability			12.00	3.00	1.00	4.00	(press F9 to recalc)
C = Confidence			12.00	3.00	4.50	1.36	Note that typical Beta's are
			12.00	3.00	1.37	2.75	between 1 and 3 for wear out

calculated

Note: Statistics does not prove anything If 12 parts are tested to 1 life then we have failed to disprove the hypothesis. If 12 parts are tested to 1 life and at least one fails then we reject the hypothesis that the population has a 90% reliability.

N1 = Number of samples for testing to one life
Nx = Number of samples to testing for x lives
Beta = Weibull slope
x = Number of lives to run Nx to achieve the same R and C of N1 to one life
Assumption: parts are tested to N or Nx lives with no failures

Figure 2-4: Reduced sample size with increased testing time based on assumed probability distributions. A fully functional spreadsheet is available on the CD-ROM.

On the companion CD-ROM, you will find a spreadsheet called *RC and Wiebull*. This is a simple tool that I put together to quickly play with reliability, confidence, sample size and Wiebull distributions. For a test that is based on *N* number of parts to one life, there is a known reliability and confidence. This means that I can choose two of the three variables and calculate the third. The spreadsheet is set up to allow any two of the numbers to be put in and the third is then calculated. Next, the number of samples needed for X lives given a Weibull shape factor of beta is calculated. I usually pick two Weibull numbers at either end of the expected range and then pick one randomly in the middle. This allows the user to press F9 (recalculate) repeatedly to see how many lives for a given sample size is needed to achieve the same confidence and reliability target as the one life test.

Twelve parts for one life corresponds to three parts for three lives for a beta of 1.26. If the shape factor is lower than 1.26, then the test could result in a false sense of security, while being higher than 1.26 could result in an over-designed part.

How do you know that the population of parts has a Wiebull distribution with a shape factor of 1.26? You would have to test a considerable number of parts TO FAILURE to determine this. Assuming a value of 1.26 would most likely result in a conservatively tested, over-designed part (since the part is most likely to have a Wiebull shape factor higher than 1.26).

Now sum it all up. With the assumptions in the vibration profile, the assumptions in the thermal profile and the assumptions in the statistics (everything driven towards the more conservative test), what will the test results mean? That the sample of the population tested supports the hypothesis that there is a low probability of failure under the test conditions before one life.

Consider the different business models from Chapter 1 and the types of decisions that individuals may be trying to make.

Top-down, commodity, price point-driven commodity:

- Can the product constantly perform for the warranty period?
- What will the warranty rate and cost be?
- Will the part work in the field on a consistent basis?
- Will the part work as well in a new market?
- Will the part fail in a dangerous way?

Cross-functional, entrepreneurial team:

- Does the product function?
- What are the basic failure mechanisms of the product?

Cross-functional, price point-driven commodity:

- Can the product be made cheaper and still perform as well or better?
- What will the warranty rate and cost be?
- What are the basic failure mechanisms of the product?
- Will the part work consistently in the field?

- Will the part work as well in a new market?

- Will the part fail in a dangerous way?

- Are failure modes a function of design or production?

I won't bias your opinion by tagging the questions as answerable or not with the results of the test described previously. But there are some things to consider when looking at the questions and evaluating the usefulness of the test for the answer.

1) Can the question be phrased as a hypothesis that is addressed by the test?

2) Are the assumptions embedded in the question controlled in the test? (If the new market is a factor, is it present and controlled?)

I think you'll quickly find that the test is well conceived, and will give accurate results on the hypothesis of the test as it was designed. But when applied to the real questions, distilling an answer from the results is not easy.

Now, consider a couple of human factors: You're the entry level validation engineer who just finished running this test because it was the same test your boss ran. You are called in to the design review meeting of a cross functional team. The team must issue a report to the engineering and production manager on the state of the product and its readiness for production. You are the one who conducted the testing, and everyone is asking you questions like the ones above. Are you going to say, "I'm sorry, this test only determines a narrow statistical question and can't really be applied to the questions you're considering?" No, the usual safe response is, "The product passed the validation test without any problems." Everyone is happy, they have a warm feeling of having addressed the questions, and off we go to production. Just make sure you change jobs sometime in the next six months.

One of the common frustrations in engineering and production is getting warranty failures back from the field that do not match anything produced during the testing. The reality is that the previously described test is destined to miss large issues precisely because it is precise.

Huh?

Yes, because the test is designed to be repeatable and controlled. The stresses are applied in a narrow range, assumed to be the "worst case." The conditions are narrowly controlled, with little variation. Suppose that a thermal gradient of 3 degrees per minute caused a catastrophic failure? What if sitting for a 4-day weekend in a particular vent position caused the unit to jam? What if it's not the vibration spectrum that causes a problem, but a particular time sequence of vibration loading? Because the test is designed to be highly repeatable and precisely applied, the issues at the edges of the test plan and the nonintuitive combinations of stress are never seen.

Joe Morrill, a colleague of mine for a time at Entela, used to work for General Motors at the Millford Proving grounds. He tells the story of a particular suspension component that was failing in the field, but passing the traditional track testing. It took awhile, but in the end the culprit was the highway expansion joints at 60 mph. The joints produced a very low-level vibration, but the impulses came at regular intervals that excited a particular vibration mode shape that caused a bushing failure.

Key facts going forward:

1) Changes in business practices have not been reflected in validation changes.

2) A precise test may not be useful.

3) Historical momentum impedes the adoption of new methods.

What Not To Know

"A stitch in time saves nine."

Let's examine the timeframe of bringing a project to market from different perspectives and examine the impact on the value of the information versus its timeliness.

To do this with some level of quantification, a measure on the "value" of the information must be made. This can be straightforward. For example, if you knew what a lottery number was going to be, the value of the information would equal the jackpot. However, in most cases the value of information is subjective, and in some cases, knowing the information precludes an objective quantification of its value.

For example, knowing that a particular design feature will break under service conditions in the field is valuable. A design fix can be made, and the problem can be avoided. But how valuable is it? The only way to know for sure is to leave the design alone and allow it to fail in the field and then measure the economic impact. The value of the information is in the cost *avoided*. Proving what this cost is means proving a negative...this design change prevented a serious warranty problem associated with the design feature.

Have you heard the one about clapping keeping the lions away? A man standing on the corner clapping furiously is asked by a police officer what he is doing. "Keeping lions away" was his answer. Looking around bewildered, the officer said, "I don't see any lions around." To which the man replied, "Works well, doesn't it?"

Knowing a design feature will break in the field and cause some warranty rate does not provide the information needed to quantify the value of the information. On the other hand, knowing the economic impact of a particular failure mode without knowing the design feature or failure mechanism would quantify the value of the information, but not provide the information to fix the problem.

As seen in Chapter 2, different tests can provide different types of information. Knowing the economic impact of a particular failure mode without knowing the failure mechanism or how to fix it is not very useful. Knowing a failure mechanism and how to fix it without knowing the economic cost of fixing it can be frustrating. And finally, the economic cost of quantifying the economic value of fixing a failure mode is usually cost prohibitive.

What is the value of a piece of information during a project?

Information types:

- Feasibility
- Physical properties
- Failure mechanisms
- Warranty rate
- Nominal operating conditions
- Maximum and minimum operating conditions
- Storage conditions
- Potential failure modes
- Component failure rate
- Mean time to failure
- Mean time between failures
- System reliability

Project stages:

- Research

- Feasibility

- Development/design

- Design validation

- Production ramp

- Production validation

- Production

- Service

Quantifying the value of each type of information at each step of a project is not possible. But providing a relative gauge of the value of the information *is* possible.

For example, the value of knowing at the feasibility stage that a product is NOT feasible is worth much more than discovering it at production. And discovering at production that the product is not feasible is worth more than discovering it after three months of production. This is an extreme case, and unlikely to occur in real life. It is far more likely that a product is discovered to have serious design flaws during production ramp or later. The value of knowing about serious design flaws during design validation instead of during production ramp or production validation is obvious, but not always quantifiable—unless you didn't know.

The value of a piece of information is only known when it comes too late. However, the cost of getting the information can be anticipated. The cost of determining if a material will work (not just prove feasible) in the design is very expensive during the feasibility stage. Because there are no tools made, no prototypes—everything would be from scratch. The cost of determining if a material will work at production or field use is fairly low, because lots of parts are available. Two curves can then be used for the ranking of cost to find information vs. the cost of not knowing. The "sweet spot" is where the sum of the two curves is lowest. (See Figure 3-1.)

Since the only way to examine the value of a piece of information at a particular time is to examine the cost of NOT having the piece of information, let's examine three scenarios:

1) A key physical property is wrong.

2) A primary failure mode of a product.

3) The mean time to failure.

In each case, we will consider the potential economic impact of NOT knowing this information at different levels of development.

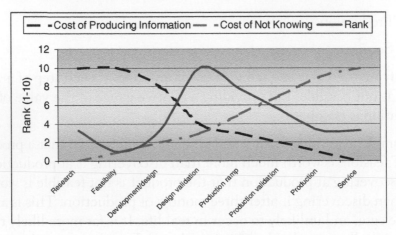

	Feasibility	Physical properties	Failure mechanisms	Warranty rate	Nominal operating conditions	Maximum and minimum operating conditions	Storage conditions	Potential failure modes	Component failure rate	Mean time to failure
Research	9	4	5	0	2	2	1	4	1	0
Feasibility	10	5	7	0	5	5	2	7	3	1
Development /design	8	8	10	2	10	10	4	10	10	2
Design validation	6	10	10	4	10	10	5	10	10	3
Production ramp	3	6	8	6	10	10	7	10	7	4
Production validation	2	3	5	8	10	10	8	10	4	5
Production	1	2	2	9	10	10	10	10	3	6
Service	0	1	1	10	10	10	10	10	2	10

Figure 3.1: Ranking on the value of information at a particular level of product development. Ten being the most valuable, zero being the least. Different job responsibilities would find different pieces of information of different values at various times during product development.

Scenario One: A key physical property is wrong.

Suppose that a modulus of elasticity on a structural plastic component is assumed to be 2000 MPa and designed accordingly. Now consider the consequence of not discovering that the material only has a 500 MPa modulus under certain temperature conditions at different times during the development process.

Research: During research, this information would be of some interest, but no economic impact would have been incurred by not knowing yet.

Feasibility: This may impact the feasibility of a product, 1/4th the stiffness is a big difference. However, choosing a different material at this point has very little economic impact provided the material costs are comparable.

Development/design: Getting into the design phase and then discovering the stiffness problem would be annoying, but the economic impact would only be slightly greater than if the discovery was made during feasibility.

Design validation: Having the design process mostly complete and then discovering the stiffness problem would cause some significant redesign and material selection rework.

Production ramp: Tools have been cut, cost of discovery now could be devastating if the replacement material could not be used in the same tooling. Plus, add in all the design and validation rework.

Production validation: Same as production ramp, but add any more time lost and missed production start. Even more cost if there are penalties for stopping another production line the part supplies.

Production: Production lines are being shut down, product deliveries are being missed. This is now a major emergency.

Service: Product has made it out the door and is in the hands of the customer. Add up everything previous and then calculate lost customers, customer dissatisfaction and so forth. And that assumes that the stiffness problem does not cause any harm. If so, add in liability costs.

Scenario Two: A primary failure mode of a product.

A resistor in a power circuit is marginally sized for the application. The normal variation in the population is not a problem, but colder operating temperatures can push the resistance too low and result in a rectifying circuit receiving excessive current.

Research: During research, this information would be of some interest, but no economic impact would have been incurred by not knowing yet.

Feasibility: This would not affect the feasibility of the product. A different resistance value or a tighter resistance tolerance would solve the problem

Development/design: Getting into the design phase and then discovering the resistance problem would be annoying, but the economic impact would only be slightly greater than if the discovery was made during feasibility.

Design validation: Discovering the problem during design validation would not cause much more harm than during development. A few more validation tests would be necessary after the design correction was made.

Production ramp: Cost of discovery now could be significant. It would involve revalidating the design, and resetting the production.

Production validation: Same as production ramp, but add more time lost and missed production start. Even more cost if there are penalties for stopping another production line the part supplies.

Production: Now production lines are being shut down or all components are being reworked, product deliveries are being missed. This is now a major emergency.

Service: Product has made it out the door and is in the hands of the customer. Add up everything previous and calculate lost customers, customer dissatisfaction and so forth. And that assumes that the over current does not cause any harm (like a fire). If so, add in liability costs.

Scenario Three: The Mean Time to Failure (MTTF).

A serviceable product has a mean time to failure. Consider two possibilities: Low MTTF or High MTTF.

Research: During research this information would be of some interest, but no economic impact would have been incurred by not knowing. The reality is that whatever MTTF the product may exhibit during research, feasibility or development may very well change due to design or production process changes or due to end-use changes. More important, the cost of determining the MTTF would be wasted since the design will go through changes.

Feasibility: This would not affect the feasibility of the product.

Development/design: Knowing the product has a low MTTF would be useful, only if the cause of the failure was known.

Design validation: Marginal value.

Production ramp: Marginal value.

Production Validation: Marginal value.

Production: Marginal value.

Service: Knowing the MTTF (low or high) is valuable at this point so that proper service can be planned for and provided. If the MTTF is low, the cost of raising it is high.

In reading through these scenarios, two salient points should be noted: 1) The cost of responding to information goes up the later the information is learned. 2) Information has no value unless it can be responded to (see Chapter 1).

The other fact that comes out of this discussion is the problem of not knowing or noninformation. If you recall Hiesenburg's uncertainty principle: Measuring the energy of a particle precludes accurately knowing its position, however, measuring its position precludes accurately knowing its energy. This principle was observed in particle theory because the smallest unit that could be used to measure either position or energy was a photon, and the photon would either change the particle's position or its energy depending on how it was used to measure the state of the particle. The quantity NOT measured becomes unknown or noninformation.

In a macro sense, the same thing happens in development. If a piece of information is gained during development, then the consequence of NOT knowing the information until service cannot be found. This is not to say that there is not a definite consequence of discovering the information too late, but that knowing the information early causes those involved in making decisions about the project to act differently. Like the particles in Hiesenburg's uncertainty principle: when and how the information is acquired precludes the ability to know other pieces of information.

This is not to say that the energy of a particle for which a position has been measured does not exist, it simply cannot be known with certainty. Likewise, the consequences of NOT knowing a serious design flaw at the development stage exists, but is not knowable with any certainty. The energy in the particle can be extrapolated from past measurements, estimated and theorized. Likewise, the consequences of NOT knowing a piece of information that is known can be extrapolated from past experience, estimated and theorized.

Dealing with noninformation in an efficient development plan is essential. An organization must have the self-discipline to make good decisions about what information to gain AND WHAT INFORMATION TO NOT KNOW. (See Figure 3-2.)

What information to know and what information to not know is affected by the business structure. A design and manufacturing firm that separates the design function from the production and service efforts will make decisions about what information to know and what information to not know within the divisions. In other words, the design team will tend to choose to know information about the design characteristics, and less about the warranty or serviceability. Measuring and demonstrating that the design is feasible and can be made reliable does not quantify what the warranty is. However, if the organization of the company dictates that the design group is doing the testing, then the

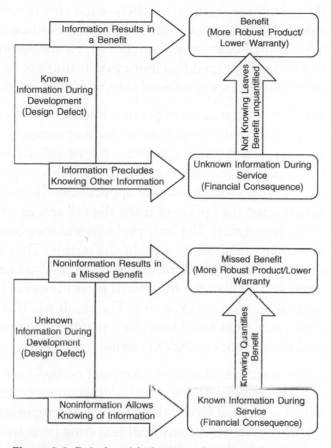

Figure 3-2: **Relationship between known and unknown information.**

testing will be conducted to focus on the design feasibility and reliability. The warranty or serviceability issues will get second seat. Reverse the situation and have the testing conducted by a manufacturing reliability supervisor who is responsible for warranty and service rates and the testing will focus on those issues. These situations force a body of noninformation, not because knowing one piece of information precludes another, but because the business unit spending the time, effort and money to get the information dictates what is known and what is marginalized.

Not only will the business organization dictate the type of testing and what information is known and not known, but the business type will also. An entrepreneurial business will choose to test the feasibility and features of the product. But spending money on features or functional testing reduces what will be spent on durability, reliability or serviceability.

The type of product being made is important also. A manufacturer of a component who sells their component to another company for assembly into a larger system will not and may not be able to test their product in the assembly. I recently consulted with one company that manufactured seat hinges for automotive applications. In discussing options for testing setup the option of using the full seat as a fixture for their product was brought up. The only problem was they could not get the seating from their clients in time to do any testing. They could not purchase after-market seating because of small changes and because the seating was often sold as all of the individual parts instead of the whole seat, therefore, making the cost excessive. The result was that the seating hinge supplier could not know how their part would perform in the full system until their client conducted testing.

The development stage also creates two kinds of noninformation. The first is simple, there are some characteristics that cannot be known about the product until final production. During development, most injected-molded dies do not reflect any final texturing (that rough surface that is given to some plastic parts like orange peel, leather and so forth). The final texturing is the last thing that is done to an injection-molding die. This means any affect of the texturing on durability or more likely interference issues with matting parts is not known during early development. The same is true of the effect of any design characteristic that cannot be realized until the final production is started.

The second type of development stage noninformation is a little more subtle. During development certain design assumptions are made, decisions about which design option to use. The cost of developing a particular design option and bring it to production often precludes switching to the other design option after tooling and production has been set. The performance, characteristics and benefits of the other design become noninformation. This is not to say that the design could not be investigated later, but an economic barrier develops the further the development goes on a different design. Once one design has been in use for a couple of years of production, the cost of switching to a different design concept becomes a serious economic challenge. This also happens on different levels, from small components to entire transportation systems. Currently, the automotive companies are trying to use some hybrid cars that have both gas fired engines and electric motors in them. They are also looking at fuel cells, natural gas, and hydrogen powered. The fuel cell cars hold the most promise for clean fuel-efficient cars, but the hybrid cars are already on the road. The hybrid cars are not necessarily better—more likely worse—but the infrastructure of gasoline fuel stations around the country make them feasible. There is no infrastructure for the fuel cell cars. The lack of infrastructure makes trying the fuel cell cars on the same scale as the hybrid cars nearly impossible. The fair comparison of fuel cell verses hybrid in the real market place becomes noninformation.

You don't know, what you don't know.

Why is noninformation important? Ignorance of your ignorance is not a good thing. Being conscious that a business structure has precluded knowing certain information is critical to making good business decisions. Suppose you have a company where the manufacturing department is in charge of the testing. The testing focuses primarily on the population's behavior and optimizing manufacturing and not much design related information. The company decides it needs to improve reliability, because its warranty rates are higher than its competition. Where to focus its efforts? Well—the information available in the records of all the testing highlight areas of strength and weakness in the manufacturing process. Great—focus on the weaknesses in manufacturing to improve the reliability. But what about the design group? There is no evidence that they have a real problem because the testing focuses on the manufacturing side. A good manager who is conscious of what he doesn't know will be asking some tough questions of the design group also.

Key facts moving forward:

1) Information is only valuable at the correct time.

2) Determining one piece of information may preclude knowing other information.

3) An efficient validation plan is as much about deciding what NOT to know as what to know.

4) The business structure, place in the supply chain, type of product and type of business dictate what will be noninformation.

4

Accelerated Testing Catalog

"The more alternatives, the more difficult the choice."

—Abbe' D'Allanival

This chapter contains a catalog of validation tools (new and old) and what kind of information they produce effectively and their production timeframe. DFMEA, Fault Tree Analysis, fully censored testing, step-stress testing, accelerated reliability, HALT, FMVT®[1] and computer modeling are summarized in this chapter. Refer to Chapters 5–11 for details on the tools.

TOOL NAME:
Design Failure Modes and Effects Analysis (DFMEA)

References for Execution

1) AIAG FMEA-3

2) SAE J1739

3) MIL-STD-1629A

4) Xfmea

Information Produced

Consolidated view of potential failure modes, their expected causes, severity, likelihood, method of detection, Detectability, corrective action and collective risk (Risk Priority Number – RPN).

---

[1] FMVT is a registered trademark of Entela, Inc.

Precision (graduation)

Generally a subjective analysis on a scale from 1 to 10 based on group input.

Accuracy (average on the mark)

Generally a subjective analysis, accuracy will depend on the participants knowledge of the product.

Repeatability

On general information, the results will be highly repeatable from group to group. Detailed information on failure modes and corrective action will be highly dependent on the design group doing the analysis.

Limitations

Highly subjective and depends heavily on the knowledge, experience and self-discipline of the group doing the analysis.

Time Frame

Can be done poorly to provide paperwork to pass an audit in a few hours. Conducted properly, it can take 5% to 10% of the labor for a development project. For example, two people working for 2 years = 2 * 40 * 52 * 2 = 8,320 hours, would need to spend between 416 and 832 hours (2 hours a week each) developing and maintaining an effective DFMEA.

Resources

For a poor job: one person.

Typical for a good job: 2–10 people, a comfortable conference room with a computer and projector. A spreadsheet or FMEA software.

TOOL NAME: Fault Tree Analysis (FTA)

References for Execution

1) IEC 60300-3-1

2) Milena Krasich, "Use of Fault Tree Analysis for Evaluation of System Reliability Improvements in Design Phase." Proceedings, Annual Reliability and Maintainability Symposium, January 2000, Los Angeles, California Attachment – Tutorial Visuals.

3) Joanne Bechta Dugan, "Fault Tree Analysis of Computer-Based Systems," 1999 Tutorial Notes, Reliability and Maintainability Symposium, Washington, D.C.

Information Produced

"The FTA allows pictorial representation of the system, its architecture and functionality, along with using Boolean algebra and the multitude of modeling schemes to best represent the system operation and interdependency of its failure modes."[2]

Precision (graduation)

The precision of an FTA can be quite good if the analysis is done in detail. If subsystems are represented as a single block in the FTA then the analysis will not be as precise.

Accuracy (average on the mark)

Accuracy is based on the potential sources of fault modeled. Most electronic components are rated well for temperature (accurate) and whether or not vibration is present. If other sources of fault to the device are not figured in, the accuracy may be off.

Repeatability

Because the FTA is a theoretical model of the potential fault paths in the device, it can be repeated exactly each time. In practice, the assumptions made in the FTA from one practitioner to another could cause some variation.

---

[2] "Fault Tree Analysis in Product Reliability Improvement," Milena Krasich, P.E., Reliability and Maintainability Symposium Tutorial, 2003.

Limitations

Most electronic components are rated for temperature and sometimes vibration. Without the knowledge of the probability of failure for components as a function of other stresses, the FTA becomes a theoretical model of failure as a function of operating temperature.

Time Frame

Depending on the complexity of the device and the skill of the practitioner, an FTA can be developed in a couple of days or a few weeks. If the probability of failure as a function of time and temperature must be determined, then the time frame will be much longer (see accelerated reliability, Chapter 8).

Resources

References for the probability of failure vs, temperature (and other stress sources, if possible) for each component. If the information is not available, then a basic testing capability including a thermal chamber and power supply may be needed to determine the relationships (see Chapter 8).

TOOL NAME: Fully Censored Testing

References for Execution

1) *Reliability and Life Testing Handbook*, Dimitri Kececioglu, PTR Prentice-Hall, Inc, 1993.

2) *Accelerated Testing: Statistical Models, Test Plans and Data Analysis*, Wayne Nelson, John Wiley & Sons, 1990.

Information Produced

Pass or fail against a desired reliability goal. The test is designed around a hypothesis: the product has a given reliability with a given confidence. The hypothesis is tested by determining how many parts should last for a given period of time at a given severity with no failures. The test is called *fully censored* because (if it passes) the test is ended before any parts fail. In other words, the time-to-failure is censored.

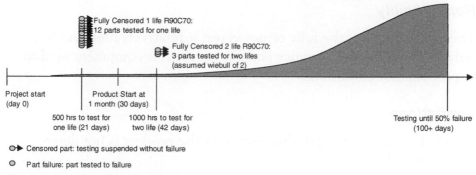

Figure 4-1: Fully censored test.

Precision (graduation)

The precision of the test depends on two key factors:

1) Number of parts tested.

2) Number of lives tested.

The greater the number of parts, the more precise (and more accurate) the test is. Increasing the number of lives decreases the number of parts needed, but introduces an assumed distribution in the population.

Accuracy (average on the mark)

The fully censored test is a pass/fail test. Therefore, the accuracy is not easily quantified. However, two factors do affect the accuracy:

1) The severity of the test relative to service conditions.

2) The tolerance on test condition replication.

One assumption of the fully censored test is that there is a "worst case" condition that can be quantified. If the severity of the test is not set correctly, the test will be inherently inaccurate. Also, if the worst case condition is not a single condition, then the accuracy of the test will be diminished for each worst case condition not included in the test. Finally, the replication in the test conditions of the target conditions will have a tolerance based on the equipment. This tolerance will also reduce the accuracy of the test.

Repeatability

The repeatability of the fully censored test is relatively high and is only degraded by the tolerance possible on the test equipment used to impose the test conditions.

Limitations

1) The actual strength distribution of the product is not determined.

2) If the part passes, there is no information to help improve the product.

3) The test can take a fairly long period of time.

4) The test can be highly misleading if the severity of the test is inaccurate relative to the service conditions.

Time Frame

Depending on the definition of "one life" for the product based on engineering analysis (see Chapter 6), the test can take a few hours to several months. Usually, the fully censored durability test will take less time for simple mechanical stress conditions (such as vibration or mechanical load), and will take much longer for thermal, chemical and some electrical loads.

Examples:

Headlight test: 1000 hours

Automotive HVAC: 36 hours

Switching relay: 72 hours

Clothes dryer: 6 months

Resources

The resources for a fully censored test depend on the service stress applied. For many items experiencing durability testing, vibration and temperature are the primary stresses. Other stresses applied may be voltage, mechanical, pneumatic, hydraulic, barometric, humidity, salt spray, chemical and radiant heat. The resources needed for a vibration and temperature depend on the accuracy used in the replication of the service conditions. Most fully censored tests use a single axis of vibration and relatively low thermal ramp rates.

TOOL NAME: Step Stress Testing

References for Execution

1) *Reliability & Life Testing Handbook*, Dimitri Kececioglu, PTR Prentice-Hall, Inc, 1993.

2) *Accelerated Testing: Statistical Models, Test Plans and Data Analysis*, Wayne Nelson, John Wiley & Sons, 1990.

3) *Accelerated Stress Testing Handbook: Guide for Achieving Quality Products*, H. Anthony Chan, John Wiley & Sons, 2001.

4) *Test Engineering: A Concise Guide to Cost-effective Design, Development and Manufacture*, Patrick D. T. O'Connor, John Wiley & Sons, 2001.

Information Produced

A step stress test starts the same way a fully censored test starts. A fixed number of parts run through one life. After one life, the stresses applied to the product are elevated in steps in order to precipitate failures. This results in a life demonstration followed by an identification of failure modes. (See Figure 4-2.)

Precision (graduation)

The precision of the test depends on three key factors:

1) Number of parts tested.

2) Number of lives tested.

3) How the stresses are increased.

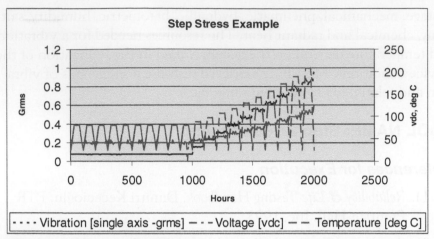

Figure 4-2. Step stress example.

The greater the number of parts, the more precise (and more accurate) the test is. Increasing the number of lives decreases the number of parts needed, but introduces an assumed distribution in the population.

Increasing the stress reduces the graduation on the time-to-failure. At low stress, the time to accumulate damage is very long and differences between times to failure will also be long. As stress is increased, the rate of damage increases, the time-to-failure decreases and the spread between times to failure decrease.

Accuracy (average on the mark)

The first part of the step stress test is a fully censored, pass/fail test. Therefore, the accuracy is not easily quantified. However, two factors do affect the accuracy:

1) The severity of the test relative to service conditions.

2) The tolerance on test condition replication.

One assumption of the fully censored test is that there is a "worst case" condition that can be quantified. If the severity of the test is not set correctly, the test will be inherently inaccurate. Also, if the worst case condition is not a single condition, then the accuracy of the test will be diminished for each worst case condition not included in the test. Finally, the replication in the test conditions of the target conditions will have a tolerance based on the equipment. This tolerance will also reduce the accuracy of the pass/fail portion of the test.

The accuracy of the second part of the test depends on how stresses are increased. If all of the stresses are stepped in a way to ensure that the rate of damage for each failure mechanism is kept in proper proportion, then the accuracy will be driven by the same issues as for the one life portion of the test. If the stresses are increased in a way that biases the test toward one failure mode, then the time-to-failure results will be inherently inaccurate.

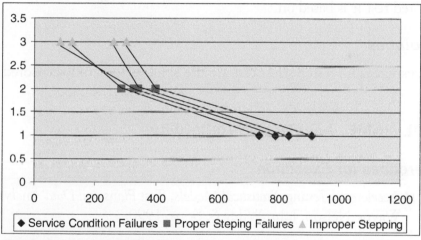

Figure 4-3: Time-to-failure in service conditions, proper stepping and improper stepping of stresses.

Repeatability

The repeatability of the step stress test is relatively high and is only degraded by the tolerance possible on the test equipment used to impose the test conditions.

Limitations

1) The actual strength distribution of the product is not determined.

2) The test can take a fairly long period of time.

3) The test can be highly misleading if the severity of the test is inaccurate relative to the service conditions.

Time Frame

Depending on the definition of "one life" for the product based on engineering analysis (see Chapter 7), the test can take a few hours to several months. Usually, the fully censored durability test will take less time for simple mechanical stress conditions (such as vibration or mechanical load), and will take much longer for thermal, chemical and some electrical loads. Generally, the steps stress test is twice as long as the fully censored test it is based on.

Resources

The resources for a step stress test are the same as for a fully censored test.

TOOL NAME: Accelerated Reliability

References for Execution

1) *Accelerated Testing: Statistical Models, Test Plans and Data Analysis*, Wayne Nelson, John Wiley & Sons, 1990.

Information Produced

Life vs. stress vs. probability. (See Figure 4-4.)

Precision (graduation)

The precision of the test depends on four key factors:

1) Number of parts tested.

2) Number of parts tested to failure and number of parts censored.

3) The number of different stress levels used.

4) The number of different stresses used.

As the number of parts increases, the precision or resolution of the results increases. However, the number of parts censored (not allowed to run to failure) may decrease the precision. The more stress levels used, the more precise the results, provided the number of samples at each level is maintained. The number of different stresses used can increase the precision, but increases the number of samples needed.

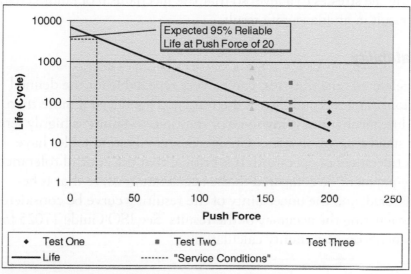

Figure 4-4: Basic accelerated reliability principles.

Accuracy (average on the mark)

The accuracy of the test depends on:

1) Math model used.

2) Number of failure mechanisms involved.

3) The stresses used and ignored.

4) The elevation of the stresses.

If the wrong math model is used to fit the data to a curve, the results can be very precise, but wrong. Choosing the correct math model is very important. Wayne Nelson's book on accelerated testing is a good reference for determining which model to use.

If there is more than one failure mechanism involved and the analysis is done under the assumption of one math model, the accuracy will likely be poor.

Using the correct stress sources is critical. If a key stress source is not used to accelerate the time-to-failure, the prediction of time-to-failure will be inaccurate.

Elevating the stresses to a level so that the mechanism of failure changes can result in an inaccurate result.

Repeatability

An accelerated reliability test can be very repeatable for the defined conditions. The repeatability of the test can be reduced if more than one failure mechanism is involved or the time-to-failure is highly sensitive to small changes in stress. Chambers and vibration tables have known tolerances of operation. It is critical that the control tolerance of the equipment be compared to the acceleration curve that is being measured, and the uncertainty of the resulting curve be considered when evaluating the accuracy of the results. See ISO Guide 17025 for requirements on uncertainty calculations.

Limitations

The biggest limitation for this test is the indeterminate amount of time necessary to execute the test. Because the most accurate results require testing to failure, the time necessary can vary widely. Running the samples for a fixed period of time can result in a more defined test, but may reduce accuracy.

Sample size is also a significant limitation. Sample sizes of three sets of eight (24 parts) or more are common. In addition, if the extrapolation to service conditions yields a tight margin relative to a target, more samples and runs may be needed in order to verify the initial results.

One stress source results in two or three coefficients in the equations. More than one stress results in exponentially more coefficients.

Time Frame

Typically, a few weeks of testing can provide the results for the higher stress levels. However, the lower stress levels may take a very long time to verify the results, especially if there is little margin between the service condition target time-to-failure and the testing results.

Resources

Resources can vary depending on the stress source being used to accelerate the time-to-failure. Often, temperature is used and an environmental chamber is all that is needed to conduct the test. Often the test setup can be more complicated, involving chamber, power supplies, data loggers, vibration equipment and so forth.

TOOL NAME: Highly Accelerated Life Testing (HALT)

References for Execution

1) *Accelerated Stress Testing Handbook*, H. Anthony Chan, John Wiley & Sons, 2001.

2) *Accelerated Reliability Engineering: HALT and HASS*, Gregg K. Hobbs, John Wiley & Sons; 1st Edition (April 10, 2000), ISBN: 047197966X.

Information Produced

Failure modes, operational limits, destruct limits. (See Figure 4-5.)

Precision (graduation)

Measuring the precision of HALT must be broken down into two distinct parts.

The precision of operational and destruct limits is affected primarily by the number of samples tested. However, the method used to determine the exact stress level of each failure does impact the precision. If a very coarse step is used, then the precision will be lower. If a finer step is used, or a binary search is used, then the precision can be much better.

For the failure modes precision is very high—for a given set of conditions the same physics of failure will often exhibit itself with a high degree of consistency. The precision can be thrown off considerably if the diagnostics and instrumentation cannot detect the failure when it occurs.

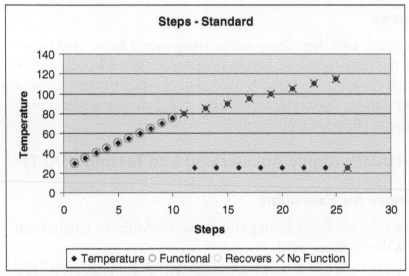

Figure 4-5: Margin discovery process.

Accuracy (average on the mark)

The accuracy of HALT is also broken down into two distinct parts.

The accuracy of operational and destruct limits is affected primarily by the coarseness of the steps used to find the limits.

The failure modes in general are fairly accurate, the main source of inaccuracy is a change in the physics of failure due to elevated stress.

Repeatability

Generally, repeatability is very good on a specific location of a specific machine. Some significant variability has been noted at different locations of air hammer tables (the most common vibration source for HALT) and between tables. Failure mode information, operational margin and destruct margins are highly repeatable for a given machine at a given location.

Thermal chambers can easily be ± 4 degrees over time and across the volume of the chamber.

Limitations

The lack of statistical measure requires a different approach to assimilating the information into an engineering organization.

The use of a large number of stress sources results in a larger sample size and a fairly long test period.

Developed for solid-state electronics. Can be used on other devices, but care must be taken to understand how and why the technique works on solid-state electronics and what must be done differently to function on other devices.

Time Frame

Setup: Typically one or two days for vibration and thermal on a powered solid-state device. For larger, more complex items involving more sources of stress, setup times can rise quickly to several weeks or months.

Testing: One day per stress plus a day for combined stresses. For a normal vibration and thermal, five days can be expected (hot, cold, ramp rate, vibration, combined). More stresses require more days.

Post Test: The main time driver on post test analysis is the failure analysis. Most failure analysis is straightforward, but some tools such as the Scanning Electron Microscope (SEM), acoustic microscopes, and infrared spectroscopy may require some days of preparation and analysis. Rule of thumb: a day to sort through every ten failures to identify the two or three that need a day or two of evaluation and occasionally one that will need a week or two.

Resources

Usually, HALT is performed using thermal and vibration on a solid-state electronic device. In this case an air hammer chamber provides the vibration and the thermal (air hammer provides high frequency vibration suited to solid-state stuff. See Chapters 9 and 10 about vibration spectrums). A power supply, diagnostic equipment for the device under test, a good technician to run everything and the design responsible engineer should all be present. Pizza is optional, but this type of test does lend itself to engineers and technicians camping out in the lab for extended hours during the few days of testing.

TOOL NAME: Failure Mode Verification Test (FMVT®)

References for Execution

1) Porter, A., "HALT to FMVT The Migration of Highly Accelerated Life Testing from Solid-State Electronics to Mechanical Systems," SAE International Congress & Exposition, March 1999, Detroit, MI, USA. Session: Accelerated Testing Conference (Part A&B).

2) Porter, A., "Design-Information Driven Testing," *Time Compression Technologies*, December 1999, Volume 4 Issue 7.

3) Porter, A., "What Process For What Part?" SAE 2000 World Congress, March 2000, Detroit, MI, USA. Session: Accelerated Testing Conference (Part A&B).

4) Porter, A., "FMVT and Accelerated Testing Overview" IEEE, AST-Symposium, Boston, 2000.

5) Porter, A., "Speed To Market and Product Durability...Could Become the Victim of Business Reality." International Appliance, KY, 2000.

6) Porter, A., "Does High Reliability Equal Zero Defects?" NPSS Workshop, Boxboro, MA, 2000.

7) Porter, A., "Using Accelerated Testing Methods to Improve Electronics Design," *Compliance Engineering*, Annual Reference Guide, 2001.

8) Porter, A., "How Do You Logically Set Up Stress Sources For An Accelerated Stress Test?" *Sound and Vibration Magazine*, April, 2001.

9) Porter, A., "Don't Stress Over Accelerated Testing," *Sound and Vibration Magazine*, May 2001.

10) Porter, A., "Quality Begins with Good Data," *Quality Magazine*, May 2001.

11) Porter, A., "Accelerated Testing Mistakes," *Evaluation Engineering*, May 2001.

12) Porter, A., "Optimizing Frequency and Identifying Failure Modes" North Atlantic Testing Workshop, IEEE, Gloucester, MA, 2001.

13) Porter, A., "Success Through Failure: Overcoming Two Major Hurdles In Accelerated Testing To Failure," *Test Engineering & Management*, August/September 2001.

14) Porter, A., "Automotive Validation has a mental condition," *Test Engineering & Management*, October/November 2002

15) Porter, A., "Life Estimating Techniques For Failure Mode Identification Testing Methods," SAE Congress 2002-01-1174.

16) LOP-01, 02, 03, 04, 05 Entela Standard for FMVT Execution.

17) Patents.

Information Produced

Failure modes, failure mode progression, design maturity, technological limit.

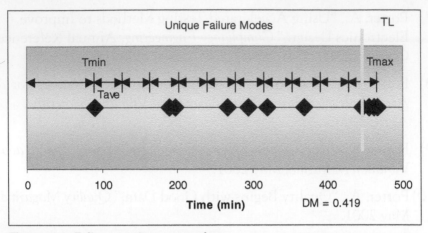

Figure 4-6: Failure mode progression.

Precision (graduation)

Overall, the precision of the failure mode progression is very accurate. The precision of the failure mode progression (and the ranking of failures) is affected by the graduation of the stresses and the effectiveness of the diagnostics. If a failure mode occurs but is not found immediately then the recorded time-to-failure will introduce an error into the design maturity measure, the ranking of the failures and possibly even the order of the failures. This in turn affects the technological limit and possibly the decisions that are made.

The design maturity measure is an order of magnitude measure and represents the approximate potential for improvement. In other words, a DM < 0.1 is considered a good part, 0.1 < DM < 1.0 is a fair part that may have some warranty issues, while a DM > 1 will generally be poor.

Accuracy (average on the mark)

The failure modes in general are fairly accurate, the main source of inaccuracy is a change in the physic of failure due to elevated stress.

The design maturity is an order of magnitude measure but is fairly accurate on average.

Repeatability

Generally, repeatability is very good. Caution should be taken on variations in vibration level from table to table and from location to location on tables. FMVT has been conducted on a wide range of vibration equipment. Failure mode information, design maturity, and technological limits are highly repeatable.

Thermal chambers can easily be ± 4 degrees over time and across the volume of the chamber.

Limitations

The lack of statistical measure requires a different approach to assimilating the information into an engineering organization. Using the design maturity and the technological limit can help in this effort.

On large systems the fixturing can become quite complex.

Time Frame

Setup: Typically one or two weeks for a moderate test with 5 to 15 stress sources. For larger, more complex items involving more sources of stress setup times can rise quickly to several weeks or months.

Testing: Typically 10 to 20 hours of testing time plus operational checks between levels. The operational checks will drive the variation in actual testing time. Some products will require a longer period of time due to their internal operational cycles.

Post Test: The main time driver on post test analysis is the failure analysis. Most failure analysis is straightforward, but some tools such as Scanning Electron Microscope (SEM), acoustic microscopes, and infrared spectroscopy may require some days of prep and analysis. Rule of thumb: a day to sort through every ten failures to identify the two or three that need a day or two of evaluation and occasionally one that will need a week or two.

Resources

Usually FMVT is performed using a thermal and vibration chamber combined with mechanical and electrical cycling, additional contaminants, product loading and so forth (anything that will break the product). The vibration table is chosen based on the resonance properties of the product under test. See Chapters 9 and 10 about vibration spectrums. A power supply, diagnostic equipment for the device under test, a good technician to run everything and the design responsible engineer should all be present. Pizza is optional, but this type of test does lend itself to engineers and technicians camping out in the lab for extended hours during the few days of testing.

TOOL NAME: Computer Modeling

References for Execution

1) *Applied Finite Element Analysis for Engineers*, Frank L. Stasa, Francis Lee Stasa, International Thomson Publishing, 1995.

2) Algor FEA Software

3) SDRC

4) Pro E/Mechanica

5) Ansys

Information Produced

Computer modeling techniques can provide a wide-range of theoretical information including stress, strain, deflection, fluid flow, heat transfer (conductive and radiated), circuit analysis, dynamic analysis and more.

Precision (graduation)

A computer model can be incredibly precise. The limits on precision are governed only by the size and length of run time desired.

Accuracy (average on the mark)

Accuracy in computer modeling is governed by two primary elements. First, most computer models use some sort of linear or polynomial piece-wise approximation. These models can be extremely accurate as the mesh or time steps are refined. Second, the boundary conditions, material properties and other assumptions introduce an inherent source of inaccuracy to the model. The degree of inaccuracy will depend on how accurate the material properties are and how many simplifying assumptions are made.

Repeatability

A given computer, with a given software package will be nearly exactly repeatable. Solving the same problem using different algorithms can be nonrepeatable, depending on the accuracy of the algorithms.

Limitations

The more complex and dynamic the computer model, the longer it takes to set up and the more computer horsepower is needed. Multiple stress interactions can be difficult, and computer models should be verified with physical testing.

Time Frame

A simple 2-D model can take a few milliseconds to run (longer to hit the enter key and save the results to the hard drive than to conduct the actual run).

A complex 3-D model with dynamic interaction (front-end automobile collision) can take several days to run on a fast computer.

Resources

Simple computer modeling can be conducted with a programming language and a computer. More complex models require software, and computers with plenty of memory and hard drive space.

5

Design Failure Mode Effects Analysis (DFMEA)

"Those who cannot remember the past are condemned to repeat it."

— George Santayana, *The Life of Reason*, Volume 1, 1905.

"Why you may take the most gallant sailor, the most intrepid airman, and the most audacious soldier, put them at a table together-and what do you get? The sum of their fears."

—Winston Churchill

Design Failure Mode Effects Analysis (DFMEA) is the disciplined analysis of potential failures in the design. The DFMEA is a team effort usually conducted by a facilitator who collects the team's input and guides the processes. When conducted properly, the process will identify the key functional items, potential failure modes, their root causes and any corrective action. The process leads to a better design and can help guide the testing and validation process. If used correctly, it can provide context to the data that physical testing will produce so the behavior of the company can be influenced (see Key fact #1 from Chapter 1).

References:

- SAE – J1739 v003 Aug 2002.
- AIAG FMEA-3
- MIL-STD-1629A
- Xfmea

Basic DFMEA

The basic parts of the DFMEA:

Functional Item: The functional feature or design feature from the Bill of Material (BOM).

Potential Failure Mode: Key word is *potential*. What failure modes could the feature experience? Source for this is engineering experience, warranty data and pure imagination.

Potential Effect(s) of Failure: What are the results of the failure on the function or behavior of the product? Often the failure itself is not visible, but the functional effect will be apparent. For example, a sealed bearing may have excessive wear, but the failure is not visible externally. The effect of increased friction is a slowing of the motor or an increase in the power draw. The effects are key to designing instrumentation and operational checks for tests to verify the existence of failure modes.

Severity: How bad are the consequences of the failure?

Criticality: How critical to the function of the device is the failure mode?

Potential Causes: The key word here is *mechanism*. What can break the product?

Occurrence: What is the likelihood of failure?

Controls: What is the current design effort to prevent the design from failing?

Detectability: How well can the failure mode be detected if it exists?

RPN: Risk Priority Number is the multiple of all four ratings: Severity × Criticality × Occurrence × Detectability.

Responsibility: Who will take responsibility for implementing the recommended action?

Target Date: When will the recommended action be completed?

The DFMEA process should start at the conception of the design process and should be kept current throughout the process and lead into the Process Failure Mode Effects Analysis (PFMEA) and provide the foundation for any follow–on development process. DFMEA's can take a significant amount of effort. One DFMEA team that I participated in for a tier one automotive supplier met for 4 hours every other week. The team had eight permanent members and called in between 5–10 additional people as the need arose. This process continued for over two years. That comes to over 1,664 hours sitting in a room and discussing potential failures and what to do about them. The time investment needed to do a good job with the DFMEA leads to a couple of scenarios.

1) Photocopy and forget.
   Copy a DFMEA of a previous project, make a few change and file it. This will satisfy the quality auditors but provides no real engineering value.

2) Strong start and revival.
   Start the DFMEA at the initiation of the project only to be sidetracked by the details of getting all the work done. Then at the end, revive the document and try to update it to reflect all of the design changes and actions taken.

3) End game scramble.
   Put off doing the DFMEA until the final quality submission is due, and then scramble to generate an 800-page spreadsheet to reflect 18 months of designs, revisions, and lessons learned (remember cramming for finals?).

On the one hand, a DFMEA does not require any prototypes or testing time and can provide a key method of preserving "corporate knowledge." However, the amount of time required and the pressures of product development often compromise the process. I have participated on a few very good DFMEA's. Unfortunately, this seems to be the exception and not the rule.

Examine the way a DFMEA is supposed to work:

Initiation: A DFMEA should start as soon as the design development process starts. At this stage, the details of specific design features may not be available and the potential failure modes will be naturally broad. Going through the disciplined process of capturing all potential failure modes at this stage will help to drive the development more efficiently. The DFMEA at this point can be used to develop the *general* outline of the validation plan.

Design Iteration: As the design is iterated, the DFMEA should be kept current. This will include adding details as design features are developed and changing details as design changes are made. The DFMEA at this point should be used to begin planning the *details* of the validation plan.

Design Validation: At this point the design should be nearly complete. The DFMEA should reflect all the details of the design and the corresponding potential failure modes. Most of the changes to the DFMEA should now be reflecting the closing of recommended actions. The validation plan should reflect checks on *all* the key assumptions in the DFMEA.

Production Validation: The DFMEA should still reflect minor changes implemented to improve or correct production problems. Since many DFMEA's are used as the basis for the next project, this step is critical. Validation should reflect checks on the changes.

Production Run: The DFMEA should still reflect minor changes implemented to improve or correct production problems. Since many DFMEA's are used as the basis for the next project, this step is critical. Validation should reflect checks on the changes.

You can see from this flow that a DFMEA can become the central focus for the knowledge of a product. The key is to make the validation and engineering both intimately connected to the DFMEA. One way to do this is to fill out the DFMEA from the middle out. Start at the Mechanism of Failure and work in both directions: go right to fill out the Current Design Controls and the Recommended Action; go left to fill out the Potential Effects of Failure, Potential Failure Mode and the Functional Item.

The reason for doing this is three-fold:

1) The mechanisms of failure are limited. If you have ever filled out a DFMEA or looked at a well filled out DFMEA, you have invariably noticed that the mechanisms of failure are highly repetitive. If Temperature is a mechanism of failure for one

functional item it will be for several others also. The Functional Items, however, are theoretically infinite: It depends on what detail you want to go to. You can go down to each component in a system, or you could go down to every fastener and every wire.

2) Design engineers design product to handle a task in an environment. Therefore telling an engineer that high temperature is likely to cause a particular failure mode on a given feature is more relevant then simply saying that a feature has a potential failure mode. The key information is not the individual feature, but the mechanism of failure first, then the failure modes and finally all of the features that are affected. Why? Regardless of the features used in the design, the mechanism will still be present, so the designer must work with the mechanism of the failure. The features can change and the resulting failure mode of the feature due to the mechanism can change, but the mechanism will always be there.

3) Validation testing is based on the mechanism, not the feature.

What this means is that starting a DFMEA by listing the potential mechanisms of failure gives the whole process a jump-start and provides the working list for the common language between the designers and the testers: the mechanisms of failure.

Why is it important to have the DFMEA fit with the validation plan? Remember that the DFMEA is based on the individual's assumptions of what the potential failures are. The design will be developed based on these assumptions. The purpose of validation is to validate that the design will behave in the real world as well as it does in the designer's mind. (See Figure 5-1.)

Hypothesis and the DFMEA

If you consider the DFMEA from the viewpoint of the scientific method, a couple of key points about the structure and use of the DFMEA become obvious. The Failure Mode is actually the Null Hypothesis. In other words, the potential failure mode is what the design assumes will not happen. What is not clear in the standard DFMEA format is how the accuracy of the assumptions (Hypothesis/Null Hypothesis) is tested. The recommended action should contain the method by which the hypothesis and null hypothesis is verified. The second thing that is missing from the DFMEA is what to do if the hypothesis is false.

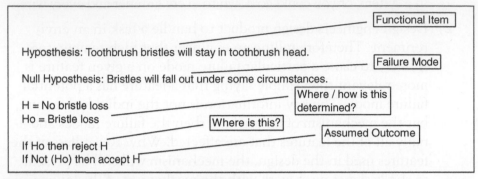

Figure 5-1: Hypothesis model of a DFMEA

To make the DFMEA more effective and tied more closely to the validation plan, add a column called *Contingency* next to the Current Design Controls column. This clearly shows that if the hypothesis is correct, the Current Design Controls will remain; if the null hypothesis is true, then the Contingency will be tried. Clearly declaring the contingency allows the development timeline to reflect the actual decision based on the information.

Function Item	Potential Failure Mode	Potential Effect(s) of failure	Severity	Criticality	Potential Causes(s) / Mechanism(s) of Failure	Occurrence	Current Design Controls	Detectability	RPN	Recommended Action	Respons-ibility	Target Date

Figure 5-2: Standard DFMEA format.

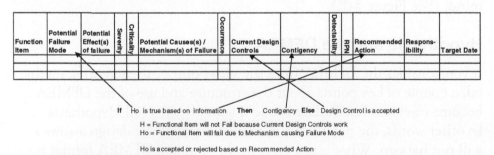

Function Item	Potential Failure Mode	Potential Effect(s) of failure	Severity	Criticality	Potential Causes(s) / Mechanism(s) of Failure	Occurrence	Current Design Controls	Contigency	Detectability	RPN	Recommended Action	Respons-ibility	Target Date

If Ho is true based on information **Then** Contigency **Else** Design Control is accepted

H = Functional Item will not Fail because Current Design Controls work
Ho = Functional Item will fail due to Mechanism causing Failure Mode

Ho is accepted or rejected based on Recommended Action

Figure 5-3: Modified DFMEA format reflects a complete scientific method.

Using the modified format, the Recommended Action contains the key steps for the validation plan. The actions that determine the accuracy of the assumptions define the test plan. The recommended actions may also contain other action items relevant to design changes and other issues. However, every assumption should have objective evidence clearly documented to support the use of the design control or the implementation of any contingency.

Consider an example. We will look at a familiar item: a toothbrush.

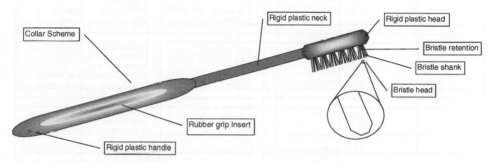

Figure 5-4: Toothbrush design.

First thing to consider is what all of the design characteristics are—the bill of material may be a good start, but keep in mind that some design features go beyond just the material it is made out of. For example, the head, handle and neck would all be molded out of one piece of plastic—but they should be considered separately to ensure the different potential failures are documented.

The example DFMEA is included on the companion CD-ROM.

Now consider just the handle. One logical failure mode to be considered is that the insert area causes the handle to split. The effect of this could be the loss of the rubber insert, separation of neck and cutting or hurting the consumer's hand. There are many potential causes: impact, thermal cycle, chemical attack/material incompatibility, fatigue and sharp radius.

Function Item	Potential Failure Mode	Potential Effect(s) of Failure	Severity	Criticality	Potential Causes(s) / Mechanism(s) of Failure	Occurrence	Current Design Controls Prevention	Detection
Handle	split in grip insert area	Loss of rubber grip	3		impact	8	Impact resistant plastic	FEA model of impact from 3 likely directions
			3	3	thermal cycle	8	thermal set plastic with stable material properties from -30 deg C to 100 deg C	Thermal cycle testing
			3		chemical attack/material incompatibility	8	chemically inert plastic to mild alkali's and acids	chemical exposures
			3		fatigue	8	FEA model of maximum loading: stress must be below 1/3 material yield.	load testing to verify models
			3		Sharp radius	5	All design radius must be greater than 1 mm	
	separation of neck		6	8				
	cutting or hurting consumers hand		8	8				

Figure 5-5: Handle portion of an example DFMEA. See the CD-ROM for details.

The critical part of this example for the purposes of this book is the detection methods. Notice that for each potential failure, there is an effect, a mechanism and a corresponding method of detection. The method of detection should reflect the suspected mechanism and take advantage of the potential effect to design a test that will impose the mechanism and monitor for the effect. For example, the loss of rubber due to impact can be tested by imparting an impact and monitoring for rubber loss. Notice that a DFMEA could lead to a very large number of discrete tests. Just a brief look at *one* failure in *one* design feature results in four tests. Two of the tests are relatively quick (FEA model, load testing), and two of them could take a significant amount of time (thermal cycle, chemical exposure).

This is one of the big downfalls of the DFMEA—if it is conducted properly, it will result in a very exhaustive list of discrete testing.

Current Design Controls Prevention	Detection	Detectability	RPN	Recommended Action	Responsibility	Target Date
Impact resistant plastic	FEA model of impact from 3 likely directions	3	216			
thermal set plastic with stable material properties from -30 deg C to 100 deg C	Thermal cycle testing	3	216			
chemically inert plastic to mild alkali's and acids	chemical exposures	2	144			
FEA model of maximum loading: stress must be below 1/3 material yield.	load testing to verify models	1	72			
All design radius must bo groator then 1mm		2	90			

Figure 5-6: DFMEA – handle design controls and detection.

Consider the mechanisms of failure again. Notice that the mechanisms of failure and the resulting testing are all assumed to be single mechanism failures—in other words, only one source of damage. Because of the linear, left-to-right thinking inherent in the DFMEA format, the notion of multiple stress sources is not immediately obvious. Failures due to multiple stresses and unique stress combinations are not obviously arrived at from the format. However, if the four mechanisms of failure were combined into a single test (impacts during thermal cycling, chemical exposure, with discrete loads), then a more comprehensive test that includes complex interactions of the stresses could be found.[1] In later chapters, the DFMEA will be referred to for mechanisms of failure and effects of failure in developing stress conditions and instrumentation for other accelerated tests.

---

[1] Porter, A., "Using DMFEA to Drive Accelerated Testing," SAE International Congress & Exposition, March 1999, Detroit, MI, USA. Session: Accelerated Testing Conference (Part A&B).

Key facts going forward:

1) The DFMEA structure is a useful bookkeeping tool, not an end unto itself.

2) The DFMEA structure can limit the weak-minded to a linear, left-to-right, single mechanism train of thought that misses more complex interactions.

3) Be sure that the DFMEA captures a complete hypothesis and contingency tied to the full validation and engineering effort. (Implement Key fact #1 from Chapter 1.)

Fully Censored Testing

> *There are lies,*
> *there are bold-faced lies,*
> *and then there are statistics.*

—paraphrased from Mark Twain

A fully censored test is the type of test that is most often seen in "standardized" testing schemes for validating designs and processes. A set number of parts exposed to a defined environment for a set period of time. If no parts fail, then the parts pass. If any parts fail before the end of the defined time, the parts fail.

Where does this test come from, why is it used, when is it useful from the business/legal standpoint, and how is it abused?

First, consider the ideal test. Enough parts tested to failure, to demonstrate, with some confidence, the assumed reliability of the product.

With this testing, two things happen. First, the sample size is large. Second, the time needed to test is large. This is because modern products are very reliable with failure rates below 1%. With failure rates this low, the sample size to demonstrate the reliability with any confidence is large.

For example, with a 1% failure rate and a desired confidence of 90%, the test must demonstrate a 99% reliability with a 90% confidence. The sample size depends on the standard deviation in the population. Generally, a set of samples (say 12 parts) is run to failure. With the results in-hand, the mean time-to-failure, the 1% failure rate and the standard deviation is calculated. If the results are sufficient to demonstrate the 1% (or less) failure rate, then the testing is done. If the data indicates a nominal 1% failure rate for one life, but the confidence is not sufficient (standard deviation too high, and sample size too low), then more samples will need to be run to achieve the confidence. The confidence interval about the mean (or about the 1% failure rate) is proportional to the standard deviation and inversely proportional to the root of the sample size. This means that as the standard deviation goes up, the sample size must go up faster. Double the standard deviation and the sample size must go up by four to achieve the same confidence.

Now ask several questions about this test. How long will it take? What information will be gained? Can the timing of the information be adjusted to maximize the value of the information?

We don't know.
Demonstration of reliability.
No.

Finally, what behavior will be changed by the results? The only behavior that can be changed is whether to change the design or not. The test is not designed to provide information on *what* to change. For completeness, consider the complete derivation of the fully censored test. We start with the basic equation and assumptions:

Probability density of failure at *x*:

$$f(x) = [cx^{(c-1)} / b^c] \exp[-(x / b)^c]$$

With the assumptions being:

1) Continuous variable (no discontinuities).

2) Homogeneous population.

3) Population is large enough to simulate a continuous variable.

Given the basic equation, a test can be determined to measure the shape of the distribution.

This requires monitoring the full population until all have failed. For reasons of time and resources, this method is not practical. Instead, a sample of the population is taken and the distribution is estimated by testing the samples to failure. However, even this test can take too long and the time-to-failure is indeterminant. Making business plans around an indeterminate time frame is difficult. This leads to one more compromise in the test: Truncate the test to a fixed period of time. The shortest period of time possible is the time expected for a given number of samples to demonstrate life with no failures.

	Full Population (Field)	Sample	Truncated Demonstration
Number of Units	100,000	100	12
Time to test (hrs of operation)	6552	6552	327.6
	To reach warranty, Approximately 10,000 to 30,000 hrs to reach 50% failure		To demonstrate reliability
Test Conditions	Field Use (6 hrs a day / days a week average)	Simulated Field Conditions (24 hrs a day)	Simulated worst 5% of field conditions (24 hrs a day)
Calendar Time (days)	1092	274	13
Data Out	Number of parts failed at any given time	Number of parts failed at any given time	Pass = no parts failed Failed = at least one part failed
Information	Actual Warranty Period Failure rate	Estimated Probability Distribution	Reliability Demonstration (90R70C)

Figure 6-1: Example of a field conditions, a sampled testing and a truncated demonstration.

Effects of sample size on information gained. Notice a couple of things about the resolution of the information gained as sample size drops. With 100,000 samples, it is possible to measure a 0.0013% rate of failure. With 100 samples, only a 1.25% failure rate can be measured, and with 12 parts a 10% failure rate can be measured (all assuming the same confidence).

This leads to an inherent monetary property of statistics based reliability demonstration. The greater the statistical reliability that must be demonstrated, the greater the number of parts and the length of the test. In other words, using statistics means the greater reliability costs more to demonstrate. This leads to the first real pressure on reliability demonstration. The immediate financial incentive is to reduce sample size and testing time. This provides immediate bottom line cost savings. The resulting degradation in information is a longer term cost driver to the bottom line that is not easily traced.

	Full Population (Field)	Sample	Truncated Demonstration	Truncated Demonstration
Number of Units	100,000	100	12	2
Time to test (hrs of operation)	6552	6552	327.6	65.52
	To reach warranty, Approximately 10,000 to 30,000 hrs to reach 50% failure		To demonstrate reliability	To demonstrate reliability
Test Conditions	Field Use (6 hrs a day 7 days a week average)	Simulated Field Conditions (24 hrs a day)	Simulated worst 5% of field conditions (24 hrs a day)	Simulated worst 1% of field conditions (24 hrs a day)
Calendar Time (days)	1092	274	13	2
Data Out	Number of parts failed at any given time	Number of parts failed at any given time	Pass = no parts failed Failed = at least one part failed	Pass = no parts failed Failed = at least one part failed
Information	Actual Warranty Period Failure rate	Estimated Probability Distribution	Reliability Demonstration (90R70C)	Reliability Demonstration (70R50C)

Resolution	0.0013%			
Confidence	72%	72%	72%	51%

Cost

unit = $10	$ 1,000,000	$ 1,000	$ 120	$ 20
Calendar day = $1,000	$ 1,092,000	$ 274,000	$ 13,000	$ 2,000
Laboratory time = $2,000	$ -	$ 548,000	$ 26,000	$ 4,000
Total Cost	$ 2,092,000	$ 823,000	$ 39,120	$ 6,020

Figure 6-2: Cost comparison of sample size and confidence.

As you can see from this table, the cost and time to do a full 100,000 unit field test is out of the question. Using 100 samples in a simulated environment is also very expensive. Conducting 12 samples to one life, which provides a 90% reliability with a 70% confidence is a reasonable cost, but the 2 parts 1 life 70% reliability 50% confidence test is over $33,000 cheaper and is finished in two days. So why not run the 2-sample test? Hold that thought.

Representative

In order for a test to be accurate, the conditions most be reproduced that the part will see in the field. This is the source of a huge amount of the cost involved in conducting testing. The field testing is often not practical on a large scale. There are companies such as AMES research that provide the automotive industry with fleet testing services that drive instrumented automobiles through a wide-range of driving conditions all over the country. This is a powerful and useful research tool, but it does take considerable time and effort. Even though the fleet testing is very close to normal service conditions, it is not exact. The cars are instrumented, driven continuously by shifts of drivers and being constantly inspected and repaired.

Testing that is done in the laboratory is done under simulated conditions. The simulation, by its nature, has some error in it. Consider temperature. Most chambers are thermocouple-controlled. Thermocouples are typically accurate to 2–4 °C. The chamber itself will have a temperature variation (properly run laboratories map and document the temperature variations within the chamber) of several degrees in small chambers and some times much more in larger chambers. This means that a test protocol that calls for simulating an 80 °C condition can be off by several degrees due to the thermocouple. If there are several samples in the chamber, the individual samples may see temperatures differing from 2–20 degrees or more depending on the size of the chamber, the circulation in the chamber, the heat load from the test parts, the quality of the insulation in the chamber and several other factors.

A product being tested at 80 °C because it represents the 95th percentile condition for the product can be tested in a shorter amount of time because the damage to the product is more severe than the product will see from the full range of temperatures. However, if the temperature that the individual part under test sees is low by 10 °C, then the testing would result in an overly optimistic evaluation of the product's performance. A higher temperature could cause a part to fail when it is just fine.

All simulated conditions have an uncertainty to their measure and to their application to the product. In addition, the target that is set (80 °C as a 95th percentile temperature) may be wrong. A long list of factors can cause the field measurements used to target environmental conditions in the laboratory to be wrong. Poor field data, field instrumentation error (the same 2–4 °C thermocouple error occurs in the field) and the time lag between measuring field conditions for development and the time the product hits the market are just a few of the reasons.

The tolerances on six axis vibration data simulation are between 5%–10%. Then take into account that the vibration data is collected from test tracks and sampled road data which has an inherent uncertainty. Some estimates are that the error between real life and the vibration simulated in the laboratory is as high as 40%.

So what do these errors mean for reliability measurement and demonstration? Well, if you're trying to measure the nominal 50% failure rate the effect is relatively small, since a 10% change in the stress level around the nominal does not affect the rate of damage to the population much. (See Figure 6-3.)

A 10% change in the stress level around the 95th percentile stress level is quite large. The variations in simulation directly impact the ability of reliability tests to be conducted in a shorter period of time on fewer samples at elevated conditions.

Homogeneous

Statistical math models operate on the premise of a continuous variable. A continuous variable is a measure or quantity that can be *any* value in a range. The voltage measured on the terminals of direct cur-

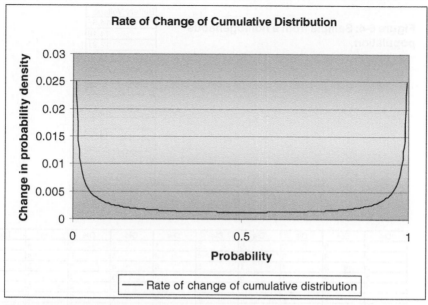

Figure 6-3: Rate of change of a cumulative distribution. Note the drastic rate of change at the extremes of the distribution.

rent batteries is a continuous variable. This means that measuring the voltage can result in *any* reading. When statistics is applied to a population, the assumption is made that a very large number of homogeneous items makes up the population. (See Figure 6-4.)

This is done because a DISCRETE variable can be approximated by a continuous variable when the number of items in the discrete population is large. A discrete variable is a characteristic that has a finite number of values. The roll of a dice is a discrete variable. A standard six-sided die will have a value of 1, 2, 3, 4, 5 or 6 showing on the top face when it is sitting on a flat surface. The ideal average of the values from rolling a die (whose sides are equally likely to come up) will be (1 + 2 + 3 + 4 + 5 + 6) / 6 = 3.5. Because the value of a die is discrete, the number of rolls that is made makes a big difference in what the MEA-SURED average actually is. Rolling the die only once will result in an average as low as 1 and as high as 6. But rolling the die 10 times results in a value more likely to be close to 3.5. (See Figure 6-5.)

Figure 6-4: Sample from a homogeneous population.

Battery Voltage
14.65
14.18
12.91
13.78
14.98
14.75
14.72
13.41
13.67
13.78
14.62
13.35

Maximum	14.98
Minimum	12.91
Average	14.06666667
Standard Deviation	0.673637795

	Die	Die	Die	Die	Die	Die	Die	Die	Die	Die
	4	5	2	5	2	1	3	5	5	2
	2	6	5	6	4	2	3	5	4	4
	2	3	4	1	5	6	6	5	2	4
	2	4	1	5	1	3	6	1	2	3
	3	6	3	3	2	4	5	4	6	2
	1	5	2	5	5	6	4	2	2	4
	3	4	1	1	5	2	6	4	1	5
	5	6	1	4	6	5	5	6	5	5
	4	6	5	5	2	4	3	2	4	5
	5	6	6	1	3	2	4	3	4	6
Average	3.1	5.1	3	3.6	3.5	3.5	4.5	3.7	3.5	4

Figure 6-5: Rolling a six-sided die ten times.

	Die	Die	Die	Die	Die	Die	Die	Die	Die	Die
	5	5	3	4	3	3	3	1	2	5
	6	3	6	5	6	2	5	1	4	2
	4	5	6	6	3	6	2	6	3	1
	1	2	5	2	2	3	5	4	5	2
	6	3	3	2	5	4	1	3	3	1
	1	1	6	5	3	2	5	2	1	3
	3	3	6	3	4	5	4	1	5	5
	5	3	4	1	3	5	4	6	3	5
	2	4	6	4	5	3	1	3	6	4
	5	5	3	4	4	3	6	1	3	5
	4	2	5	4	1	4	6	4	2	2
	6	1			3	1	2	6		1
	5		5	4		6	2		1	5
		4	3	5	2		3	2	6	2
	1	5			6	3	1	3		1
	3		6	1	2	6	2	2	3	2
	4	1	6	5	5		2	6	3	2
	2	2	2	5	5	6	4	1	6	4
	2	6	3	3	2	2	2	6	1	3
	4	4	5	5	5	6	3	3	6	6
	4	3	3	5	1	2	2	3	6	5
Average	3.55	3.65	3.81	3.5	3.39	3.77	3.57	3.36	3.38	3.36

Figure 6-6: Rolling a six-sided die 100 times.

Rolling the Die 100 Times

The average of 100 samples of a discrete variable approximate the behavior as if the value of the die could be *any* value between 1 and 6. This is called *The Law of Large Numbers.* (See Figure 6-6.)

When examining the time-to-failure of a product in the field, the time-to-failure is a continuous variable. However, the failure rate at a particular point in time is a DISCRETE variable. In other words, if there are only 20 units in the field, the actual failure rate can only be 0%, 5%, 10%,…, 80%, 85%, 90%, 95% or 100%. In most manufacturing situations, the expected field population is very large and approximating the reliability using a continuous variable is accurate.

Figure 6-7: Nonhomogeneous variable.

Battery Voltage	Die
14.65	6
14.18	3
12.91	4
13.78	4
14.98	4
14.75	5
14.72	4
13.41	4
13.67	2
13.78	4
14.62	2
10.05	2

Average	14.06666667	3.666667

Average	8.866666667

However, the other assumption in the approximation of the DISCRETE variable as a continuous variable is HOMOGENEOUS. Homogeneity means that all of the items included in the population are essentially the same—all battery voltages or die rolls. If the variable being measured is not homogeneous, the measure may be distorted.

An actual product in the field is usually made up of a variety of components, technologies and assemblies. The time-to-failure in the field will be the superposition of several different failure mechanisms. Just as the overall average of the battery voltage and die value gives an overall average that is meaningless, understanding the time-to-failure of the product would require measuring the time-to-failure for each failure separately. However, for reliability demonstrations of a given number of parts to a given period of time, only the time to first failure is found.

The graphs in Figures 6-8 and 6-9 show the probability of failure from each of four failure modes. The overall probability is then plotted against a math model fitted to the data. The math model is based on the assumption of homogeneity. The significant discrepancy between the math model and the overall probability of failure is due to the inability of a single continuous variable math model to accurately reflect a nonhomogeneous population.

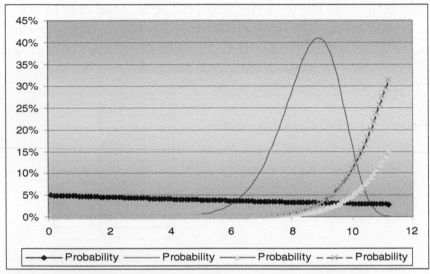

Figure 6-8: Probability plot of four different failure mechanisms.

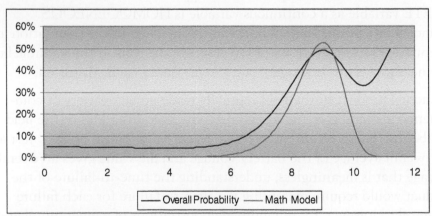

Figure 6-9: Math model vs. overall probability.

Now consider the sampling of such a population. Twelve parts tested for one life (3 years). What reliability can be demonstrated? The 90R70C (90% reliability, 70% confidence) demonstration assumes a homogeneous population, but the failures are not in any way homogeneous. The 12 parts tested are not likely to distinguish the 5% failure rate, so the parts will pass and the warranty rate will be significant.

When to Use It?

Given all of these drawbacks, why would anyone use a fully censored test?

Pros:

1) Clear, definitive pass/fail criteria.

2) Finite period of time.

3) Highly repeatable.

4) Results are well accepted and perceived as understandable.

Cons:

1) Many hidden assumptions.

2) Can provide a false sense of security.

3) If the part passes, there is no information that can guide the improvement of the product.

So when should this test be used? When the information needed is the demonstration of a MINIMUM reliability. If a more precise reliability is needed, an uncensored test should be conducted to determine an actual distribution or an accelerated reliability test should be conducted to determine the relationship between stress and life.

When considering the use of any statistical test, fully censored, uncensored or otherwise, avoid the temptation to use a statistical method simply because it has always been used. How will you use the information that you get out of the test? Will a statistical measure of a product's reliability for a given period of time answer the question you need to answer?

Take this scenario:

A tier two automotive supplier is producing an audio amplifier that will be sold to a tier one system integrator to be assembled into a door module to be sold to a North American transplant of an Asian automaker (no one said the global market place was simple). The automaker requires a demonstration of the system reliability from the tier one supplier. The tier one supplier figures out that there are seven major systems in the door: sheet metal, latch and handles, glass system, hinges, seals, amplifier and speaker. Therefore, the tier one manufacturer takes the dictated reliability goal for the system and figures out what each major component must do to ensure a reliable overall system.

Suppose that the overall system had to be 95% reliable. That would mean that the unreliability of each major component could not drive the unreliability of the full system over 5%. A straight probability calculation would mean that the 0.95 reliability = $X1 * X2 * X3 * X4 * X5 * X6 * X7$. Where X is the reliability of each major component. If each major component is held to the same standard, then $X^7 = 0.95$ or $X = (0.95)^{(1/7)} = 0.9927$. The tier one supplier could simply require each component supplier to demonstrate a 0.9927% reliability for their component.

Do you see a problem with this? How many audio amplifiers should be tested to one life to demonstrate 0.9927% reliability with a 70% confidence? 164. If the tier two suppliers conduct this test will it mean that the full system will be 95% reliable? No, for two reasons. Every statistical test has an uncertainty to it. But more importantly, the real system does not follow the underlying assumptions of reliability that were made in dividing up the unreliability to each of the subcomponents. Namely, the statistical model does not take into account the interaction of the individual components, but assumes an idealized serial system. This just means that the door fails if any one component fails, but the door also fails if any interaction between the components fail. Only the individual failures are reflected in the statistical model.

So what value is testing 164 amplifiers to demonstrate a 99.27% reliability? Assuming that there are no interaction failures between the components, then the system should have a 95% reliability (with a 70% confidence). It should be noted that in the automotive industry, the interface between components is one of the most problem-riddled areas. An electrical connector not only represents a cross between electrical, chemical and mechanical engineering, but also usually represents the boundary between responsibilities of individual suppliers.

A better approach may be to commit to the 95% system reliability and then design tests to ensure that the causes of unreliability are identified and removed from the components, their interactions and the system. Then demonstrate a 95% reliability of the full system. Three parts to three lives with an assumed 1.5 Weibull slope.

Key facts going forward:

1) Statistics is a powerful tool for quantifying probabilities.

2) A probability is not always the key piece of information needed by the project.

3) Generating data for statistical analysis often requires large sample sizes and long testing times for durability data.

4) Be sure that the assumptions embedded in the statistical method match the reality.

7

Step Stress Testing

Step with care and great tact
And remember that Life's a Great Balancing Act
Just never forget to be dexterous and deft
And never mix up your right foot with your left.

—Dr. Seuss, *Oh, the Places You'll Go.*

Step stress testing is a compromise between the fully censored testing in Chapter 5 and the need for real engineering information required by a modern quality and development system. The information that can be gleaned from this test, as well as ways the test can be abused and used correctly will be detailed.

A step stress test does not appear at first glance to be an "accelerated test." In fact, a typical step stress test will take twice as long as a corresponding conventional fully censored reliability test from Chapter 6. The step stress test is an accelerated test, not because it gets the same information in a shorter period of time, but because it gets several times the information in only twice the time.

At the end of Chapter 6, we established that a successful fully censored test produces little information when the product passes. Because none of the parts are supposed to fail, very little information is available to help improve the product, or to evaluate the margin with which the product passed. The step stress test is the answer to that dilemma.

CAUTION: If you have been using a standard fully censored test approach and you have reached this chapter, you may be very tempted to implement the techniques outlined here and forget the rest of the book...don't. Step stress testing is very powerful and is a huge step forward over the conventional fully censored test, and in some situations it is the correct test...but not in all. You've been forewarned...read the whole book.

A basic step stress test starts with a fully censored reliability demonstration. In the example given a one life test consists of vibration applied in a single axis at a nominal 0.08 grms, voltage applied at a nominal 42 Vdc with occasional drop outs, sags and swells, and temperature applied between 25 °C and 80 °C in ten thermal cycles. Twelve parts would be tested to the one life (1000 hours) to demonstrate a 90% reliability with a 70% confidence. After the 1000 hours, the stresses are amplified a small amount every 10% additional life. This amplification continues through 2 lives (1000 hours in this example). (See Figure 7-1.)

If the product fails during the stepping portion of the test, the part should be repaired or the component that failed should be replaced and the test continues.

The results of this test may look like the following:

12 parts passed to one life
90% reliable with a 70% confidence
Failures during stepping

Step	Time [hrs]	Failure
3	1235	Front left bracket fatigued at corner on unit one
3	1279	Power connector fatigued at + wire unit five
5	1405	Right rear bracket fatigued at corner on unit three
5	1454	Front left bracket fatigued at corner on unit eleven
6	1528	Right rear bracket fatigued at corner on unit seven
6	1582	Amp A2 burned out on unit six
7	1646	Front left bracket fatigued at corner on unit nine
8	1732	Right rear bracket fatigued at corner on unit eight
9	1881	Front left bracket fatigued at corner on unit five
10	1987	Amp A2 burned out on unit seven

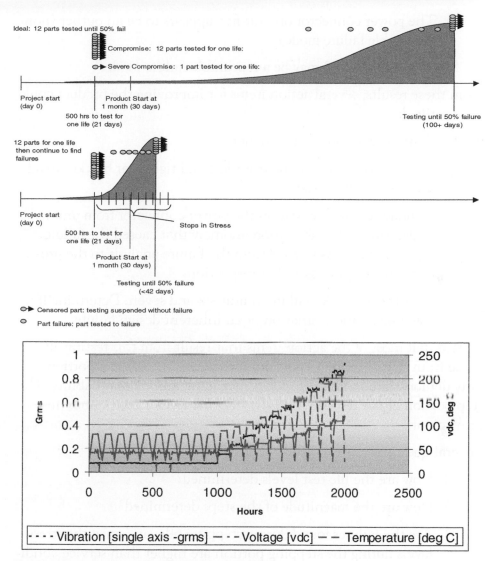

Figure 7-1: Example of a step stress example.

From this data, several important things can be concluded:

1) The products meet some minimum level of reliability.

2) The product survives more than 20% beyond the service life.

3) The front left bracket and the right rear bracket have a repeatable fatigue failure mode.

4) The power connector on unit five appears to be an outlier (non-repeatable failure mode).

5) The amplifier A2 may be a repeatable failure mode.

From these results, several action items for improving the product could be identified:

1) Start production of the product.

2) Address the design of the front left and right rear bracket to address the fatigue issue.

3) Conduct failure analysis on the power connector from unit five to determine the fabrication anomaly that caused the failure. Apply the lessons learned from the failure analysis to the production line process and/or inspection.

4) Investigate the A2 failure in unit six and seven. Determine if it was a fabrication variation or an inherent design failure.

Notice that most of the action items that result from this testing all come from the STEPPING portion of the test. The life test portion only demonstrated that the product met some minimum reliability. The step portion of the test provided the information for continuous improvement.

Several questions can be asked about this test method:

1) How are the life test levels determined?

2) How are the magnitude of the steps determined?

3) Are the failures found during stepping relevant since the stress levels during the stepping portion are higher than service conditions?

4) Is there a point at which the failures are irrelevant?

Life Test Stresses and Levels

The life test stress levels are determined just as they are for the fully censored test discussed in Chapter 6. The sources of damage to the product that are expected in the field are determined. The expected levels of the stress are determined, and the amount of time at the stress levels are determined. Usually, the one life test is based on some expected maximum stress condition and the amount of time that will be spent at the condition. For more details see Chapter 6.

Stepping Magnitude

This is the most critical aspect of getting the step stress test correct. As such, the stepping magnitude is the source of many mistakes.

Consider first the case of the fatigue properties of steel. Steel follows a known stress vs. life curve (sometimes called an S-N curve). A known increase in the stress imposed on the steel during the cycling of the steel will cause a known decrease in the life of the steel (number of cycles to break). Notice that the S-N curve has two distinct linear slopes on the log scale depending on the cycle rate and stress level. This is because the physics of the low cycle-high stress fatigue is macro work hardening—plastic deformation of the metal. This is what happens if you take a metal coat hanger and bend it back and forth until it breaks (or gets hot enough to burn your fingers). The high cycle-low stress fatigue is caused by micro-crack propagation in the surface of the metal. If you reference the *Atlas of Fatigue Curves*, edited by Howard E. Boyer, ©1986 American Society of Metals, you will find many different S-N curves for steel depending on the grade and the surface preparation. For design purposes, the high cycle-low stress portion of the curve is *usually used*. (See Figure 7-2.)

In the high cycle-low stress portion of the S-N curve the rate of damage doubles for every 10% increase in stress. As long as this increase in stress remains in the micro-crack (low stress-high cycle) portion of the curve, the physics of failure will be the same (see preceding questions 3 and 4). For this reason, when a step stress test is being conducted on a metal product where the primary failure mechanism is high cycle fatigue, the logical step in the stress is 10%. Usually this is accomplished through a 10% increase in the LOAD (for example vibration), but a 10% increase in the vibration level may or may not cause a 10% increase in the stress. (See Figure 7-3.)

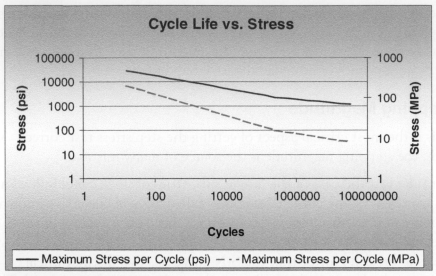

Figure 7-2: Steel Fatigue Curve.

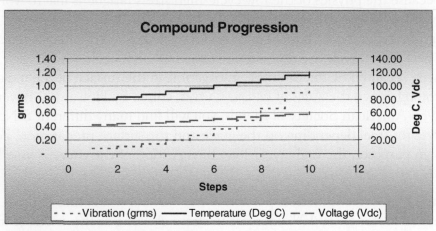

Figure 7-3: Compound progression.

Unfortunately, the 10% step rule is only appropriate for steel. Other metals have different curves, and other materials may be more sensitive to other sources of damage. A 10% increase in temperature is not a good idea as a rule (10% Rankine, Kelvin, Centigrade or Fahrenheit?). Likewise, a 10% increase in voltage may or may not have the desired result. (See Figure 7-4.)

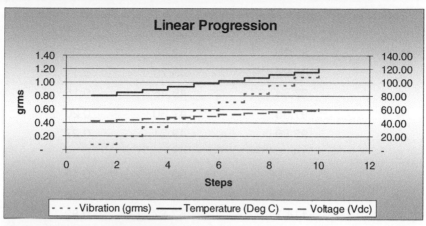

Figure 7-4: Linear progression.

Instead of focusing on increasing the rate of damage on the product, using a destruct limit or change in the physics of failure as the maximum condition to be reached at the end of the stepping provides a more consistent way of determining the stress steps.

Step	Temperature (Deg C)	Voltage (Vdc)	Vibration (grms)
1	80	42	0.08
10	120	60	1.2
Linear Progression			
1	80.00	42.00	0.08
2	84.44	44.00	0.20
3	88.89	46.00	0.33
4	93.33	48.00	0.45
5	97.78	50.00	0.58
6	102.22	52.00	0.70
7	106.67	54.00	0.83
8	111.11	56.00	0.95
9	115.56	58.00	1.08
10	120.00	60.00	1.20

Compound Progression

	4.61%	4.04%	35.11%
1	80.00	42.00	0.08
2	83.69	43.70	0.11
3	87.54	45.46	0.15
4	91.58	47.30	0.20
5	95.80	49.21	0.27
6	100.21	51.20	0.36
7	104.83	53.27	0.49
8	109.66	55.43	0.66
9	114.71	57.67	0.89
10	120.00	60.00	1.20

For example, the voltage (42 Vdc), temperature (80 °C) and vibration (grms) can be examined for the product under test, and a maximum test level can be determined based on the destruct limit (the level of stress that will cause failure) or by the change in the physics of failure. So the maximum voltage is set at 60 Vdc because there is a primary component that will blow out above this voltage. If this information is not available, a quick test can be conducted on one sample to determine the maximum voltage under which the product would work. The maximum temperature is set to 120 °C because of the glass transition point of the plastic the part is made from (glass transition point is a temperature that is usually below the melt point of the plastic that causes a significant change in the properties of the plastic). The vibration level is set by taking one part and raising the vibration applied until the part fails.

The next question is whether to use a linear increase in the stress (from the one life test conditions to the maximum test conditions), or use a compound rate (like the 10% increase in stress on the steel). This decision is difficult because not all materials follow the same exponential growth pattern that steel does. For example, many temperature-dependent electrical problems follow an inverse temperature log. For this reason, it is often advisable to use the linear progression. The linear progression is easily determined by taking the maximum test level, the service test level and dividing by the number of levels minus one (max-min) / (Levels − 1). This gives a value of (120 − 80) / (10 − 1) = 40/9 = 4.44 °C for the temperature.

For the compound progression or exponential progression, the % change per level is determined by = (−1 + (max/min) ^ (1 / (levels − 1))). This gives a value of (−1 + (120 / 80) ^ (1 / (10 − 1))) = −1 + (1.5) ^ (1/9) = −1 + 1.04608 = 0.04608 = 4.6%. A spreadsheet with all of these calculations, graphs and examples is on the accompanying CD-ROM.

Business Style

Consider a couple of applications of step stress testing. Because the focus of the test is the demonstration of the one life reliability first, followed by the identification of failure modes during the stepping process, there are a couple of logical points in the supply chain and development cycle where the step stress is uniquely qualified.

Consider a tier one supply during the production validation stage. The supplier has a requirement to demonstrate through a one-life test a given reliability: 24 parts 1 life no failure for an R90C90 demonstration. Fulfilling the one life, no failures requirement does not provide the supplier with any internal information that can be used to improve the product. The life demonstration only satisfies the contractual requirement between the supplier and the purchaser. On the other hand, the supplier is required by the restrictions of the contract to conduct the one life test. In this situation, the step stress becomes a very useful tool. The supplier conducts the one life test on the 24 parts and demonstrates the life of the product. These results are supplied to the purchaser to satisfy the contract. The supplier then continues testing the parts beyond the service conditions in the step stress manner to determine failure modes relevant to the design. This failure mode information is kept by the supplier and used for continuous improvement and future design iterations. In this way, the supplier is able to balance between the demands of the OEM and still conduct the testing to failure necessary for continuous improvement.

Now consider a similar scenario. Often, a tier one supplier is required to demonstrate a given level of reliability at the start of production, but must also make a commitment to reduce the warranty rate from the start of production going forward. For example, the supplier described above may also be required to achieve a minimum level of warranty (or less than a maximum warranty rate) the first year, and then reduce the warranty rate by a given percentage (say by 20%) the next year. This ensures a commitment to continuous improvement in the quality of the product and production. A step stress testing scheme for the production validation and ongoing sampling can help. The one life demonstration still satisfies the contractual requirement of demonstrating a given level of reliability, and the identification of failure modes during the stepping portion of the test provides the information needed to implement continuous improvement in the design.

When not to use step stress: Step stress testing uses the one life demonstration at the start of the test to establish a reliability of the product during life and the stepping portion to determine failure modes. If the only information that will be used from the test is either the reliability demonstration or the failure mode information but not both (logical OR, not AND), then use a different test for better efficiency. For a reliability demonstration, use a conventional reliability test (Chapter 6) or an accelerated reliability test (Chapter 8). For the failure mode information, use an FMVT (Chapter 10) or HALT (Chapter 9). See these respective chapters for more details.

8

Trading Stress for Time

Yes, we have to divide up our time like that, between our politics and our equations. But to me our equations are far more important, for politics are only a matter of present concern. A mathematical equation stands forever.

—Albert Einstein

Accelerated reliability will be discussed in general since this is a topic that has been thoroughly treated in other books. Scholarly references will be given, as well as guidance on when it is and is not practical to use this relatively specialized test method.

Xeno's paradox

A person stands ten feet from a wall.

They move half the distance to the wall.

They again move half the distance to the wall.

They continue to move half the distance to the wall.

Will they ever reach the wall if each move is half the distance?

This little riddle is a great question to ask to find out how people think. A practical person would observe that a point would come at which the distance remaining was so small as to be negligible and the person would have reached the wall. A pure mathematician would argue that if the person was moving "exactly half way" each time then they would never reach the wall, but the distance between the individual and the wall would go asymptotically to zero. An engineer would argue that the person would be deemed to have reached the wall when the distance was smaller than the tolerance of the measurement or controls being used to establish distance.

If you consider the wall to be failure and the distance to be time-to-failure, then accelerated testing is like moving closer to the wall in order to reduce the time-to-failure. This is done by increasing the stress on the product so that the product fails faster. In a pure mathematical sense, this is very elegant. If an increase in stress of 10% causes a doubling in the rate of damage (halving the distance to the wall), then we should be able to make the time-to-failure infinitesimally smaller and get the time-to-failure very quickly. Naturally, the practical person and the engineer point out that there must be some limit. Many would argue that any stress level above the user conditions would render the time-to-failure and the failure mechanism irrelevant. However, time and experience has shown that a product can be accelerated in a quantifiable manner within reasonable bounds. Namely, the physics of failure cannot be changed, and the time-to-failure must still be measurable. In other words, we can halve the distance to the wall as long as we don't change walls and the distance is measurable.

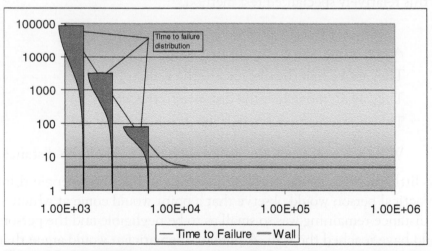

Figure 8-1: Asymptotic approach to the wall. As the stress increases, the time-to-failure (mean time-to-failure) decreases.

Basic Principles

The basic premise for accelerated reliability testing is that increasing a source of stress will decrease the average time-to-failure and the variance (or variability) of the time-to-failure. By measuring the time-to-failure at different levels of stress, the relationship between the time-to-failure and the stress level can be determined.

Description of Accelerated Reliability Method

Accelerated reliability uses a simple premise and some complex math to achieve a reliability estimate of a product for particular conditions.

As a particular source of stress is increased, time-to-failure exponentially decreases. This effect is used to design the accelerated reliability test. A simple example of this effect is the fatigue curves of steel.

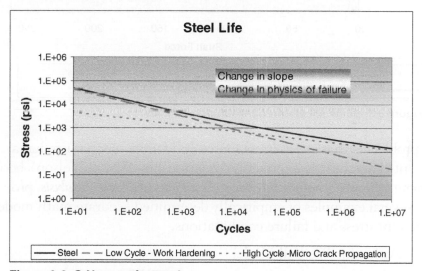

Figure 8-2: S-N curve for steel.

As the stress level rises, the time-to-failure drops exponentially. Also, the logarithmic rate of change is affected by the change in the physics of failure. Both of these effects are used in accelerated reliability.

In an accelerated reliability test, several sets of parts are tested at stress levels much higher than the expected service level. For example, 8 parts might be tested simultaneously (test one) until 4 parts failed. This would establish one point on the accelerated reliability graph. Then another 8 parts would be tested at a slightly different stress level until 4 parts failed (test two). Testing of groups of parts at different stress levels would continue until enough data was collected to extrapolate the service condition time-to-failure.

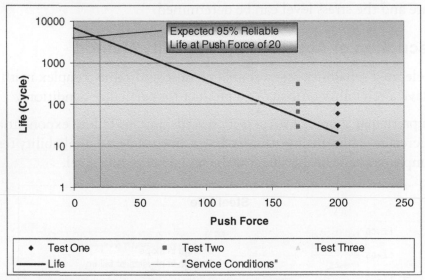

Figure 8-3: Basic accelerated reliability principles

The problem is that every stress source and failure mechanism has a different characteristic exponential relationship. Wayne Nelson's book, *Accelerated Testing: Statistical Models, Test Plans and Data Analysis*, provides several examples of empirically determined accurate math models of different stress and failure combinations.

- Arrhenius-Weibull Model
  - $F(t;T) = 1 - \exp\{-[t \exp[-\gamma 0 - (\gamma 1 \ / \ T)]]\}\beta$
- Power-Lognormal Model
  - $F(t,V) = \Phi \{[\log(t) - \mu \ (x)] \ / \ \sigma\}$
- Cox (Proportional Hazards) Model
  - $R0(t) = \exp[-\int 0t \ h0(t)dt]$

All models are from the book, *Accelerated Testing: Statistical Models, Test Plans and Data Analysis*, by Wayne Nelson, John Wiley & Sons, 1990.

Single Variable Model

Given a single stress, the accelerated testing must quantify two characteristics. First, the relationship between the stress level and the time-to-failure must be quantified. Note that time-to-failure can be a unit of time or a measure of the number of events or cycles. Second, the probability of failure as a function of time (the hazard function) must be quantified. Quantification requires determining what stress will be used and what math model will best fit. The actual testing involves exposing the product to the stress, monitoring the product for failure and measuring the time-to-failure on the product.

As an example, consider a temperature dependent failure mechanism. The Arrhenius-Weibull model has been shown to be a good approximation of the temperature effect on the behavior of the time-to-failure. Once the stress and model have been chosen, where to start? Choosing a temperature to start the test at requires several considerations.

Choosing the starting stress level:

1) The first stress level should be high enough to result in a significant number of failures. Ideally, 50% of the failures should occur in a short enough period of time for the project to move forward. Keep in mind that a couple of additional stress levels will be completed after the first stress level, and the information from the first stress level will affect how the testing proceeds. Getting the first stress level test done quickly is important.

2) The first stress level should avoid causing a change in the physics of failure. Yes, this is in direct conflict with the first point. Some obvious relationships between the stress and changes in the physics of failure can be used: glass transition temperature or melt point of plastic, thermal limits of key electronic parts, thermal expansion limits of tight tolerances and so forth. Even knowing what these characteristics are, it is possible to inadvertently cause a change in the physics of failure. In some cases, stress levels have been limited to what was thought to be an upper limit, only to find that the product was much more stable than was expected.

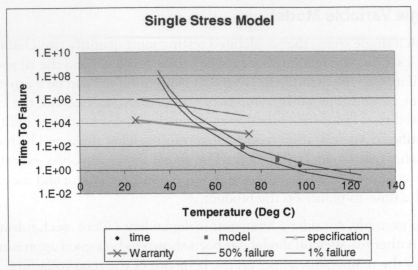

Figure 8-4: Single stress variable.

Two-Variable Model

The two-variable model must quantify six quantities. The relationship between the first stress and life, the relationship between the second stress and life, the probability of failure as a function of time for both stresses, and finally the combined effect (covariant) of the two stresses on life and probability.

$$t = ((-(LN(P))) \wedge (1/\beta_1 * (S_2/\gamma 3)))/(\exp(-\gamma 0_1 - (\gamma 1_1 * (S_2/\gamma 4)/S_1)))$$
$$+ ((-(LN(P))) \wedge (1/\beta_2))/(\exp(-\gamma 0_2 - (\gamma 1_2/S_2)))$$

t = Time

P = Probability

$\beta_{1,2}$ = Shape coefficient 1 and 2

$\gamma 0,1_{1,2}$ = Primary coefficients 1 and 2

$\gamma 3,4$ = Interaction coefficients

$S1,2$ = Stress 1 and 2

This relationship can be simplified if there is no covariant (combined effect) of the stresses. What do we mean by this? If the stresses have a covariant or combined effect above and beyond their individual effect on the life and probability, then the stress interacts to change what the simple sum of the two equations would be. If the two stresses are independent, then the effect on time-to-failure and probability by changing one stress is not affected by the magnitude of the other stress, that is to say the stresses are independent.

Following is an equation simplified for independence:

$$t = ((-(LN(P)))^{\wedge}(1/\beta_1))/(\exp(-\gamma 0_1 - (\gamma 1_1 / S_1))) + ((-(LN(P)))^{\wedge}(1/\beta_2))/(\exp(-\gamma 0_2 - (\gamma 1_2 / S_2)))$$

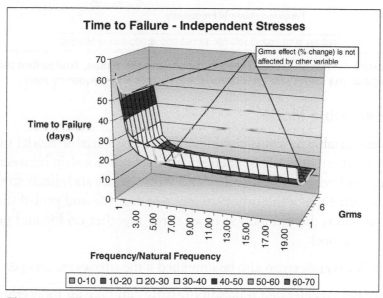

Figure 8-5: Time-to-failure curve for independent stresses.

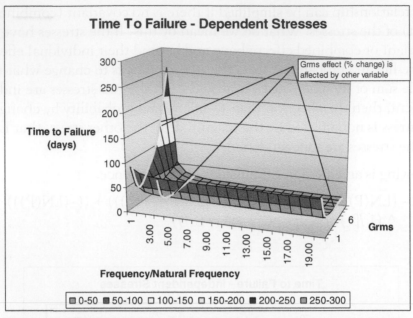

Figure 8-6: Time-to-failure curve for dependent stresses. Notice that the shape of the GRMS vs. life curve changes with the frequency ratio.

Three-Variable Model

The three-variable model is complex. The three-variable model must quantify fourteen quantities. Each stress has a relationship between life and stress and between probability and stress (that's six). Each stress interacts with each of the other stresses to affect life and probability (six more), and then the interaction of all three together on life and probability (two more).

This complex equation can also be simplified if the stresses are independent.

When to use: Accelerated reliability testing can take an indeterminate amount of time. This means that scheduling the test is more difficult than for any of the other test methods detailed in this book. Most test methods have a defined period of time or a very short test plan. The first one or two steps of the accelerated reliability (the highest stress levels) can go very quickly (1–2 days), but the final stress levels (lowest stress) can take a very long time. For this reason, a full-blown accelerated test should be used as a research and development tool to understand key characteristics of a commodity so that the design refresh cycle can move faster.

For example, consider the windings in an automotive compressor clutch. This is a magnetic clutch that is used to engage the compressor with the belt drive on the engine. The key failure mode in the device is the fatigue in the leads from the coil to the connector. This failure mode is affected by geometry and vibration—particularly vibration at and below the natural frequency of the coil. The compressor needs to be used in a variety of applications with a range of vibration magnitudes and frequencies (North American gas fired full-size sedans, European diesel powered compacts, South American SUVs). Using an accelerated reliability test to determine the life-relationship between vibration magnitude and frequency range would provide a database of knowledge that could be used as a design tool.

The test would be a two-variable accelerated reliability using vibration magnitude and frequency/natural frequency as the two stress sources. Fortunately, the two can be assumed to be independent. Three magnitudes at each of three frequencies to natural frequency ratios for a total of nine test conditions. Testing four samples at each of the nine conditions would provide 36 test points and enough data to quantify the effect of vibration magnitude and frequency on the life and probability of

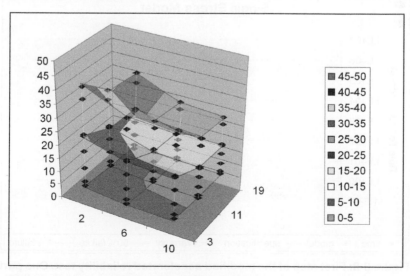

Figure 8-7: Two-variable probability surface.

failure. The nine test conditions that are most severe could take one or two days, while the least damaging may take several weeks or a couple of months. However, once the testing is done, the information provides a solid basis for adapting the compressor clutch to different conditions by designing the product to handle the different frequency ranges.

Because of the logical places to use an accelerated reliability test plan (to generate a body of data for a commodity) the test is used best by established commodity manufacturers with a diverse application environment. In other words, a one-off design would not benefit much from the test, but a family of products with a common critical design element (like a compressor clutch) would. When used properly, accelerated reliability testing produces information (instead of just data) per Key fact #1 from Chapter 1.

An accelerated reliability test should not be used during design validation testing or for noncommodity products. However, there is a modified accelerated reliability method that can be used during PV testing to shorten validation times. In this method, the maximum test level is determined and run (for example, 200 in-lbs applied to a key fastener). The resulting equation is then calculated based on what is needed for increase in life due to the reduction in stress to meet the life requirements.

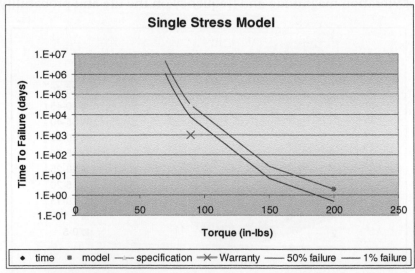

Figure 8-8: First step of the modified accelerated reliability test. One part at maximum stress.

The resulting equation is then checked at a different stress level—targeting an 8-hour, 50% failure run. If after 8 hours, no failures are found, the test level is stopped and the equation is recalculated to the minimum possible to fit the data and still meet the service condition life requirements.

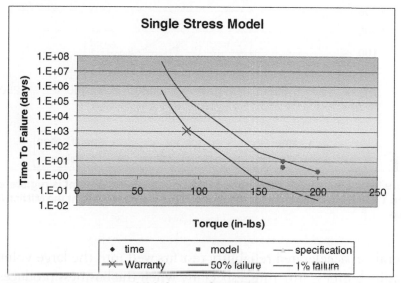

Figure 8-9: Second and third sets of samples.

Finally, one last point is checked on the equation at a lower test level (target for a few 100 hours). Again, the test is run to 50% failure or the time is exhausted. If no failures are found, then it is assumed that the equation is *better* than needed to meet the life requirement. This method manages the time slightly better than a full accelerated reliability test, but loses the actual stress vs. life equation (you know it's better than is needed, but not how good).

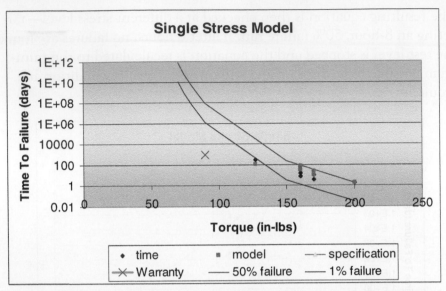

Figure 8-10: Final check. Four parts tested none failure. A point is registered at the 1% line as a reference for the model. This would represent the MINIMUM curve.

In general, an accelerated reliability test fits well into the large volume commodity business structure, especially when the business produces a family of products with common design elements used across a range of applications. In this case, the expense of producing the body of data from an accelerated reliability test can precipitate design changes and decisions throughout the product line for those common elements.

Highly Accelerated Life Testing (HALT)

Highly Accelerated Life Testing (HALT) is a test method, with its very imprecise acronym (it does not measure "Life," and there is no quantification of what "Highly" means) that has been used, abused, lauded and ridiculed, written and documented for many years. This chapter will discuss what the test method is supposed to do, and provide guidance and examples of accomplishing the test properly with references to scholarly work on doing it right. Also included are examples of how to really "mess it up."

> *"How much wood would a woodchuck chuck if a woodchuck could chuck wood?"*

The HALT process would be a good tool to determine the answer to the above tongue-twister riddle. The basic principle of highly accelerated life testing is the margin discovery process. The goal of this process is to determine the margins between the service conditions of the product and the functional limits and the destruct limits. In other words, how much wood would the woodchuck chuck, or how much temperature can the circuit board handle, or how much vibration, or how cold, or how much voltage.

HALT is a test method usually applied to solid-state electronics that determines failure modes, operational limits, and destruct limits. This test method differs significantly from reliability tests. HALT does not determine a statistical reliability or (despite its acronym) determine an estimated life. HALT applies one stress source at a time to a product at elevated levels to determine the levels at which the product stops functioning but is not destroyed (operational limit), the levels at which the product is destroyed (destruct limit) and what failure modes cause the destruction of the product.

HALT has a significant advantage over traditional reliability tests in identifying failure modes in a very short period of time. Traditional reliability tests take a long period of time (from a few days to several months). A HALT typically takes two or three days. Also, a HALT will identify several failure modes, providing significant information for the design engineers to improve the product. Typical reliability tests will provide one or two failure modes, if the product fails at all.

HALT typically uses three stress sources: temperature, vibration and electrical power. Each stress source is applied starting at some nominal level (for example, 30 °C) and is then elevated in increments until the product stops functioning. The product is then brought back to the nominal conditions to see if the product is functional. If the product is still functional, then the level at which the product stopped functioning is labeled the operational limit. The product is then subjected to levels of stress above the operational limit, returning to nominal levels each time until the product does not function. The maximum level the product experienced before failing to operate at nominal conditions is labeled the destruct limit. This process is repeated for hot temperature, cold temperature, temperature ramp rate, vibration and voltage. The process is also repeated for combined stress.

There are two significant disadvantages to HALT. Without a statistical reliability measure, the method does not fit well into the requirements for contracting between suppliers and purchasers. This relationship requires an objective measure that can be written into a contract. Some schemes have been suggested that would allow the objective measure of the relationship between the "operational limit" of the product and the service conditions. The second disadvantage to HALT is the amount of time the test method can take to address a significant number of stress sources. Each stress source tested requires about one day of testing and one or two sample products. Since HALT is usually applied to solid-state electronics, the stress sources are limited to hot, cold, ramp, vibration, and voltage. This requires 6 parts (including the combined environment test) and 4 or 5 days (8-hour days). However, applying the method to 10 or 20 stress sources increases the number of parts to 11 or 21, respectively, and the days of testing to 10 or 20, respectively.

A Typical HALT

Fixturing and Operation

Because the HALT margin discovery process is based on determining at what stress level the product stops working, it is important that the product be fixtured and instrumented to provide for quick and thorough performance evaluation. The instrumentation needed for a conventional fully censored test is usually far more spartan then for a step stress (see Chapter 7), HALT, or FMVT (Chapter 10). With a conventional fully censored test, it is assumed that the product is going to survive for the whole test period. The only piece of information that is relevant is whether the product is working at the end of the test. For this reason, the instrumentation during the test may be very simple or even nonexistent on the fully censored reliability test. With the HALT process, the goal is to determine exactly *when* and *how* the product fails.

Proper instrumentation must follow from a proper definition of what a failure is for the product and what the effects of the failure are. In Chapter 5, we discussed the failure mode effects analysis and brainstorming of the potential failures and the effects of the failure. Using the FMEA can be a good source for the definition of the failure modes and, equally important, their effects.[1]

When determining the instrumentation for the failures, the effects of the failure are often the key to properly instrument for the existence of the failure. For example, a sealed bearing has the potential failure of galling. With a sealed bearing, it is very difficult to determine the existence of galling on the sealed bearing surface. In a conventional fully censored reliability test, the part could be cut open at the end of the test. During a HALT, a less invasive way of determining when the

---

[1] Porter, A., "Using DMFEA to Drive Accelerated Testing," SAE International Congress & Exposition, March 1999, Detroit, MI, USA. Session: Accelerated Testing Conference (Part A&B).

failure occurs (and progresses) is needed. In the case of a sealed bearing, the effect of galling on the bearing performance in the system would be a logical choice. A sealed bearing in a hard drive can be instrumented for galling by monitoring the current. Since hard drive motors turn at a constant RPM, the current draw will increase if the bearing galls.

Here are some other examples of failure modes, their effect and the instrumentation:

Failure Mode	Effect	Instrumentation
Bearing galling in hard drive.	Increased current draw.	Current measurement (and voltage).
Cracked substructure in complex plastic assembly.	Shift in natural frequency of assembly.	Accelerometer at anti-node.
Seal leak in pneumatically sealed enclosure for ABS brake system.	Moisture ingress into sensitive electro-hydraulics.	Precharge enclosure with halogen gas—use halogen detector to monitor for leaks.

Notice that instrumentation of this kind requires an understanding of the whole product, not just the individual failure mode. For example, the cracked substructure in a complex plastic assembly can be detected by placing an accelerometer at the anti-node of the product. An anti-node is a place on a product that moves the most while under vibration (a node is a place that does not move while under vibration). Imagine a string that is attached at one end to a wall and you move the other end up and down. Move it at one speed and you will get one "wave" in the string, move it faster and you can get two waves. With two waves, the two "peaks" are the anti-nodes, moving the most. The point in the middle of the string that does not move is the node. (See Figure 9-1.)

When the part begins to crack, the stiffness of the product will change and the natural frequency will drop.

$$F_n = (k/m)^{1/2}$$

F_n = natural frequency

k = stiffness of the product (made up of its geometry and material properties)

m = mass

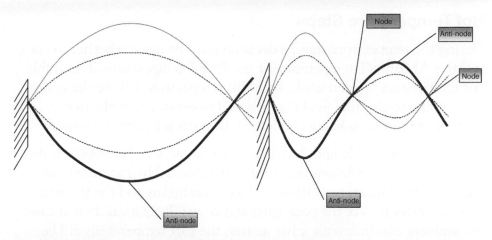

Figure 9-1: Node and anti-node illustration. The node is the point on a product (in this case, a string) that does not move due to the natural resonance of the product. It is worth noting that the anti-node of the first mode shape (left) is a node for the second mode shape.

In order to use an accelerometer to detect a crack in this manner, it is important to know where the anti-node is. This requires either a resonance survey or the use of a Finite Element Analysis (FEA) to determine the mode shapes (the shape a product takes under vibration).

If the accelerometer is not put at the anti-node, a crack could develop and not be detected. The reality is that every product has multiple mode shapes and the major mode shapes have to be considered to determine points at which the accelerometer can be placed so that it does not sit near a node and as close as possible to an anti-node. If the accelerometer sits on a node, then there is no motion from that mode shape, and detecting a change in the mode shape due to a stiffness change will be very difficult.

Once the instrumentation is determined then the testing can start.

Hot Temperature Steps

During the temperature steps, a decision must be made whether to control the AMBIENT temperature or the PART temperature. The ambient temperature is often used, since this temperature will be the easiest to correlate to the final field conditions. However, when electronic component testing is being conducted, the part temperature is used.

Why? For an assembled product like an automotive audio amplifier, the ambient service conditions are fairly well defined. The part temperatures will be a direct function of those ambient conditions and the thermal dynamic properties of the packaging and power dissipation. In this case, the ambient conditions are what matter; the part temperature will be what it will be. In the case of component testing (a capacitor, for example), the part temperature is used because the final ambient conditions of the package the capacitor will go in are not known. In fact, a given capacitor model will most likely find itself in a wide range of packages and resulting ambient conditions. The capacitor manufacturer is therefore interested in rating the capacitor for its maximum part temperature, which will be higher than the ambient temperature of the amplifier.

During the hot temperature steps, the product is held at either an ambient temperature or a part temperature (assembly or component, respectively) until the part temperature stabilizes.

Once the product stabilizes (either at the target component temperature or at a stable temperature in the target ambient temperature), the functionality of the product is checked. Note: This is a key point for making an efficient HALT test—make the performance checks fast, but thorough. It is a common mistake to plan for the time it takes to reach temperature and not plan for efficient performance evaluation. I've seen this lead to 10 days to execute a planned 2-day test because the performance tests were not planned out properly.

When the product fails the performance testing, the temperature is dropped back down to the nominal service conditions and the part is allowed to stabilize. The performance checks are rerun to determine if the part recovered. If the part does recover, then the "operational limit" has been discovered. If the part does not recover, then the "destruct limit" has been discovered.[2] The operational limit often exists for electronic components and some electromagnetic devices like motors and solenoids. Many mechanical devices will not have an operational limit (temperature at which they stop working, but recover if cooled), but will simply operate until they reach a temperature that destroys them. Some purely mechanical devices will function to temperatures well above what can be tested in a standard chamber (which can go to between 177 °C to 200 °C).

Once the operational limit has been determined, the test gets slower. The product must be brought to ever higher temperatures (at which it will not work), and then returned to nominal service conditions until a temperature is found that destroys the product (no recovery when returned to nominal temperatures). This process can take a very long time. Using a couple of extra samples can speed up the process. Establish the operational limit, then increase the temperature to a much higher temperature (for example, 50 °C) and check if the part has been destroyed. If not, go another 50 °C. Once the part reaches a 50 °C

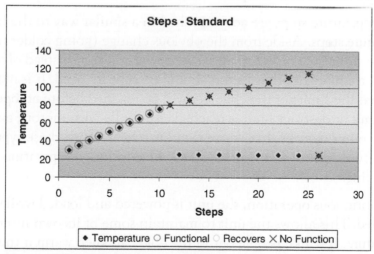

Figure 9-2: Margin discovery process.

---

[2] Gregg K. Hobbs, *Accelerated Reliability Engineering: HALT and HASS*, (John Wiley & Sons, 2000).

increment that destroys the product, use another sample and try 25 °C cooler. Continue to halve the temperature until the destruct limit is known.

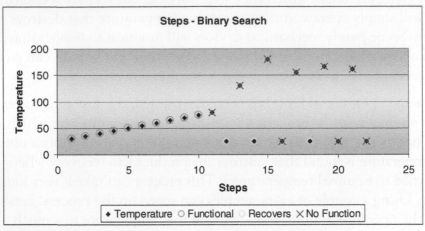

Figure 9-3: Margin discovery process using a binary search.

Cold Temperature Steps

Cold temperature steps are accomplished in a similar way to the hot temperature steps. Aside from the obvious change (going colder instead of hotter, from the coldest nominal services condition instead of the· hottest), a couple of other factors should be kept in mind. It is more likely that the limits of the chamber will be reached before an operational limit or destruct limit is reached, especially with motors, amplifiers and other devices that create their own heat. With these types of devices, two types of test may be necessary: continuous operation and duty cycling.

With continuous operation, the unit is powered and loaded while being cooled. This allows the unit to maintain some of its own internal temperature. Just like people out in the cold can keep warm if they keep moving—products that function continuously while being cooled are more likely to survive. Duty cycling the product by shutting it off—letting it cool until it has stabilized and then trying to start it up again—is a harsher test. It also takes much longer, especially for larger products. One, or the other, or both methods can be chosen based on the information needs of the project. A product like an audio amplifier that must start up on a cold winter day after sitting would benefit from the duty cycle method, while a telephone switching box that will be in an unheated shed may use the continuous method.

Because the unit may very well function all the way down to the limits of the chamber, it may be a good idea to check that temperature first, noting that if the coldest temperature destroys the part, then the sample size for the test must be increased by one. On the other hand, if it is discovered that the product survives at the coldest temperatures, then a day or two of testing is avoided.

Ramp Rates

Once the hot and cold operational and destruct limits are discovered, then the effect of thermal ramp rates can be determined. If both an upper and lower operational and destruct limit has been discovered, then a decision must be made whether to ramp between the service limits or operational limits. If there are no operational limits, avoid using the destruct limits. The purpose of the ramp portion of the test is to determine the ramp rate at which thermal expansion differences in the materials cause a failure. If the product is taken too close to the destruct limits, then the failures may not be due to the ramp effect.

Once two temperatures have been chosen to ramp between, the product is heated and cooled at a mild ramp rate while checking its functionality. Once the product has been ramped up and down at a given ramp rate without failure, the ramp rate is increased. The process is continued until the unit no longer functions. The unit is returned to nominal operating temperatures and is functionally checked again. If the unit functions, then the ramp rate is increased again. (See Figure 9-4.)

Note that ramp rates often make a bigger difference on electronic products and some functional items that are highly susceptible to the effects of different coefficients of linear thermal expansion in the materials that make up the product. The fast ramp rates cause the parts on the outer surfaces to achieve different temperatures than parts inside the product. (See Figure 9-5.)

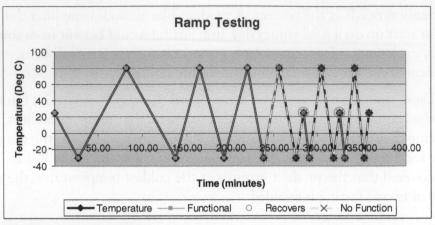

Figure 9-4: Ramp testing.

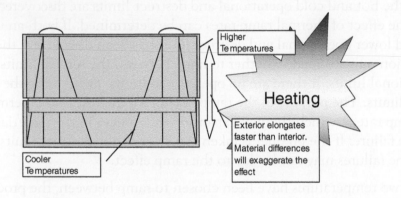

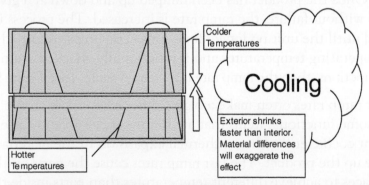

Figure 9-5: Ramp RATE and DIRECTION are both important to thermal stress.

Vibration

The vibration portion of the test is critical. Vibration is prevalent in nearly every product environment. Even a paper towel dispenser mounted on a bathroom wall sees vibration. Technically, the only environment that is devoid of vibration would be absolute zero. Everything else is vibrating.

Vibration is a very nonintuitive stress source. So let's take a little time to look at it now, and then in Chapter 10, we will explore it further.

Vibration is usually measured as an acceleration amplitude (distance per unit time squared such as m/s^2 or ft/s^2) vs. frequency. Gravity is the most prevalent form of acceleration (9.81 m/s^2) and is the subject of many high school physics questions involving throwing a ball. But vibration does not go in one direction like gravity, which pulls you towards the center of the Earth, Mars or Venus (depending on where you live). Acceleration in vibration oscillates at some rate. In fact, random vibration can cause acceleration oscillations at many rates at the same time.[3] Just like white light has a full spectrum (wavelength) of light in it, random vibration has a wide range of vibration rates. This is an important concept to grasp for proper testing using vibration. Random vibration is not a SERIES of discreet oscillations at different rates; it is all the rates in the spectrum AT THE SAME TIME. When white light is passed through a prism and is separated into its separate colors, you do not see the colors pulse or alternate because different frequencies (wavelengths) of light pass through the prism, averaging white light. They are all present at the same time.

Understanding that random vibration is a spectrum of oscillation rates or a frequency spectrum leads to a logical question. What is the frequency RANGE? Visible light has a known wavelength range, with infrared and ultraviolet beyond the visible range. Vibration has frequency ranges also. A swing on a playground oscillates at less than 1 Hz (it takes longer than 1 second to swing back and forth), while the surface-mounted chip inside your computer has a mechanical resonance (oscillation) as high as 10,000 Hz.

---

[3] J. P. Holman, *Experimental Methods for Engineers*, (McGraw-Hill Education, 1994).

Now consider the swing on the playground for a moment. To make it swing requires an input energy in the form of some type of periodic or oscillating energy source. In most cases, this is the swinger kicking their legs forward while they lean back followed by kicking their legs back while leaning forward. The other method is for someone (would this be the swingee?) to push the swinger periodically. If you have ever swung on a swing you know that you must put the energy in at the correct rate, in pace with the swing of the swing. If you try to push the swing at different times then you will not swing very far. This is critical to understanding the function of vibration in breaking a part!

What? Why?

Because, breaking a device under test requires generating STRAIN in the product. Strain is displacement in the structure. When a structure oscillates under the vibration input, it experiences strain. The amount of strain is directly related to the amplitude of the acceleration and to how well the input oscillation matches the natural oscillation of the part. Just like the swing, if you don't put the energy in at the right pace or beat, then much less strain will take place (see Figures 9-6 and 9-7). The only real difference is that swings swing at a very low frequency, and most products that we want to test oscillate at much higher frequencies. The higher frequencies contributes to the nonintuitive nature of the vibration, above about 50 Hz or so you can rarely see the motion. In Chapter 10, we will discuss the relationship between amplitude, frequency range and strain in greater detail. For now, suffice it to say that the frequency range is important.

Traditionally, HALT testing has used an air hammer table. There are various forms of these devices in existence. The basic premise is to have several pneumatic hammers repeatedly beating on the underside of a table to induce the table to resonate and generate random vibration on the top side of the table. This type of vibration is sometimes called *repeated shock* or *pseudo random*. These tables have been used because they can produce very high frequency ranges suitable for solid-state electronics.

During the vibration stage, the product is subjected to vibration at ever-increasing levels until the unit stops functioning. If the unit re-

Figure 9-6: Pushing a swing at the correct rate or pace will result in the greatest amplitude. See the CD-ROM for a movie of the swing.

Figure 9-7: Pushing the swing at the wrong time will actually slow the swing down. See the CD-ROM for a movie of the swing.

covers once vibration is removed, then the operational limit has been found. Vibration is then increased to find the destruct limit (energy level at which the unit stops functioning and does not recover). This process is much faster for vibration than for the hot or cold portions of the test because vibration takes only a few milliseconds to stabilize. (See Figure 9-8.)

Figure 9-8: Air Hammer tables (repetitive shock), Envirotronics Star44

Combined Run

Once the upper and lower operational and destruct limits have been established for temperature and the operational and destruct limits for ramp and vibration, a combined run is conducted to see if there are any failure modes that require a combination of temperature, ramp and vibration to induce failure.

Business Structures

HALT testing in the supply chain has some benefits and drawbacks. For the innovative company that is interested in understanding the functional limits of their product (under what conditions will it work), the test is extremely useful. That is to say that if the functional limits of the product is data that will affect the behavior of a sentient being, then the test is useful (see Key fact #1 from Chapter 1).

For the commodity supplier in the supply chain, there is a problem. The test finds margins and failure modes, but does not quantify how long the part will last before the failures are experienced in the field. This means the results are not useful for the contractual reliability demonstrations often required in the relationship between a purchaser and a supplier.

Manufacturers of commodity items at the component level (such as resistors and capacitors) find the test useful for determining the operational limits of their product and to determine the failure mode when it does fail. Being able to publish this information can be helpful.

When applied to design validation HALT works well, but the process is too long for production screening. HALT applied to production screening requires a slight modification. Often called Highly Accelerated Stress Screening (HASS), the production screening based on HALT uses a narrow set of conditions to precipitate failures in bad parts while leaving good parts alone. In this method, 100% of the product is screened. Alternatively, a sampled screen in which a few parts from each lot are tested to failure can be used with a full HALT.

How do you choose between 100% screen and a sampled screen? This comes down to two key factors: 1) Can the screen be engineered to precipitate failures without damaging good product? 2) Is the production process controlled enough to apply statistical sampling?

These two questions are important. If a 100% screen is applied, then the screen must not damage the good parts, but find the bad parts. But the statistical question is important also. It is more economical to use a sampled screen (few parts to test, and a more aggressive screen can be used). But a sampled screen cannot be used unless the production process is highly repeatable. When the production process is repeatable, then a statistical screen can detect drift in the quality of production. For example, sampled screen based on HALT may be used to get the following results from each lot.

Lot	Time-to-Failure
9/12/03 –1	126.4
9/12/03 –2	127.3
9/12/03 –3	125.1
9/13/03 –1	123.1
9/13/03 –2	121.1
9/13/03 –5	101.2

You can see that the time-to-failure is drifting—this would indicate a drop in the quality. Obviously, with a sampled screen, the samples can be taken to failure. With a 100% screen, the samples that are good must not be damaged. This means the screen should be tested by passing a group of samples through the screen one or more times and running the full HALT on them and compare the results to a full HALT on a set of parts that have not been screened. There should be no discernable difference.

A 100% screen should be used if: 1) the failures from the production process are random and a statistical test cannot predict the quality of the product. 2) The screen has been demonstrated to precipitate failures in bad product, while leaving good product undamaged.

A sampled screen should be used if the failures from the production process are statistically predictable due to drift in the process.

Figure 9-9: Entela FMVT "Pod".

10

Failure Mode Verification Testing (FMVT)

Don't be discouraged by a failure. It can be a positive experience. Failure is, in a sense, the highway to success, inasmuch as every discovery of what is false leads us to seek earnestly after what is true, and every fresh experience points out some form of error which we shall afterwards carefully avoid.

—John Keats

In the last chapter, Highly Accelerated Life Testing (HALT) was discussed. HALT was developed for and predominantly applied to electronics. Following is a brief discussion of how Failure Mode Verification Testing (FMVT) was developed out of HALT. This chapter will discuss what the test method does, how to set up the test method, mistakes and problems to avoid, and examples.

HALT has been around since before 1979 for testing solid-state electronics.[1] Various attempts have been made to apply HALT to mechanical testing. In 1996, I traveled to Denver to the Hobbes Engineering Symposium to learn about accelerated reliability from Wayne Nelson. There, I was also introduced to the HALT process. Working predominantly in the automotive industry, I recognized that HALT had some application, but most of the products I had to test were much more mechanical or larger electrical systems instead of solid-state electronics.

Upon returning from Denver, I began to explore why HALT worked on solid-state electronics. At the time, my primary focus had been the im-

---

[1] "Vibration Device Enters Market," *Hughes News*, September 14, 1979. US Patent 4,181,026, Abstein, Jr., et al., January 1, 1980.

plementation of Finite Element Analysis (FEA) in the lab setting. We had used FEA to model the parts we were testing (see Chapter 11). One of the first things I did was model a cup holder and apply single axis (one degree of freedom) stress vs. 6-axis (3 translations and 3 rotations) random stress to the computer model and looked at the resulting stress and mode shapes. It become obvious that the reason 6-axis random vibration worked to uncover so many key failure modes was because it had the potential to activate all the mode shapes of the product.

Not all 6-axis vibration is the same. As discussed in Chapter 9, the frequency range of the vibration matters. If the energy is not put in at the natural frequency of a particular mode shape then the energy is mostly wasted. This realization (which many other users of vibration have found[2]) led to the first key observation about HALT: "It's not the vibration, it's the stress." In other words, the vibration spectrum is not an end but a means, the means of creating random stress throughout the product. If the product experiences random stress throughout, distributed based on how the geometry and function will naturally concentrate the stress, then the weakest feature will accumulate stress damage faster than the rest of the features.

Because it is understood that the random stress in the product led to the first major divergence of HALT, the following should apply: instead of including the typical temperature, vibration, and possibly electrical stress, a thorough test on mechanical and electromechanical designs should encompass all sources of stress that can damage the product.

At the same time that this understanding was developed and first applied to a real cup holder in a physical test (sources of damage included: vibration, temperature, humidity, cycling and cup loads), a second realization was developed. The Design Failure Mode Effects Analysis (DFMEA) provided a good source for the long list of potential sources or mechanisms of damage to the part. Also, because the list of stresses

---

[2] Wayne Tustin, "A Practical Primer on Vibration Testing," *Evaluation Engineering*, November 12, 1969.

Wayne Tustin, "Using Random Vibration in Electronics Evaluation," *Evaluation Engineering*, July 8, 1978.

was now getting large (often over ten separate sources of stress), the normal process of margin discovery used in HALT was not practical. Therefore, two final developments took place: Use 10 steps with all of the stresses combined and randomized relative to each other (to keep the random stress on the product), and correlate the test to the potential failure modes in the DFMEA. These changes led to the change in test name (since it was a new test) that reflected the purpose of the test—to verify failure modes from the DFMEA-FMVT. These developments are embodied in U.S. patents: 6,035,715 – method and apparatus for optimizing the design of a product; 6,233,530 – control system for a failure mode testing system; 6,247,366 – design maturity algorithm. Entela, Inc. holds all patents; and special thanks to Mark Smith of Entela, who is coinventor on the control system patent.

In addition to the testing practices developed over this period of time, a pneumatic 6-axis vibration machine that could produce up to 4 inches of displacement and a frequency range from 5 Hz to 2500 Hz was developed. This machine (called the FMVT machine), along with air hammer tables, single-axis machines, and a 6-axis servo-hydraulic machine where used to provide the full-range of testing frequency at Entela.

Development FMVT

Failure Mode Verification Testing (FMVT) uses multiple stresses applied to the product, starting at service conditions, and then elevated to a destruct level in a stepwise fashion. The stresses are applied in a random fashion in order to maximize the number of combinations of stress that are applied to the product. The goal of the test is to find multiple failure modes, analyze the failure mode progression and determine the significant failure modes to be addressed to improve the product. The potential for improvement and the maturity of the design are also determined. (See Figure 10-1.)

In a development FMVT, the primary goals are the identification and sorting of failure modes to determine what to fix on the design to make the product more robust. FMVT drives the product toward a design where the product lasts for a long period of time and all of the dam-

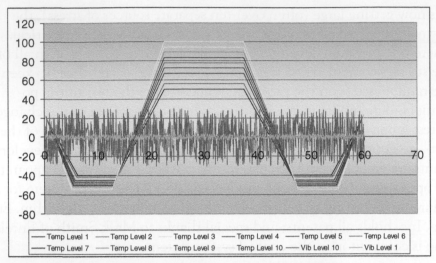

Figure 10-1: Temperature and vibration for a typical FMVT.

age is accumulated uniformly throughout the product. By driving this optimization, the FMVT results in a product that is as good as it can be for the given technology. The product can then be compared to existing designs and its reliability measured.

The goal of the test is to precipitate failure modes from all stress sources in an order that approximates their relevance. By applying all of the stresses simultaneously and elevating them from service conditions towards a destruct limit, the failures can be shown to be precipitated in approximately the order of relevance.[3]

With the FMVT, the testing is conducted on a single sample. The analysis is not statistical but is designed to check two assumptions. First, that the design is capable of producing a viable product for the environments applied. Second, that a good design and fabrication of the product would last for a long period of time under all of the stresses that it is expected to see and would accumulate stress damage throughout the product in a uniform way, so that when one feature fails, the rest of

---

[3] Porter A, "Life Estimating Techniques for Failure Mode Identification Testing Methods," SAE 2002-01-1174.

the product's features are near failure. Therefore, the hypothesis of the test is this: the product will last for a long period of time under all stress conditions and will then exhibit multiple diverse failures throughout the product. (See Figure 10-2.) The hypothesis is rejected if failures occur early or if they occur isolated in time relative to the bulk of the failures. (See Figure 10-3.)

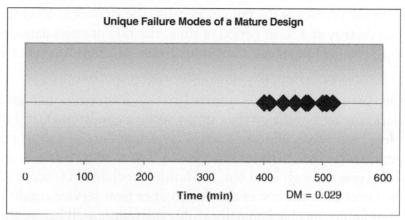

Figure 10-2: Hypothesized progression of failures.

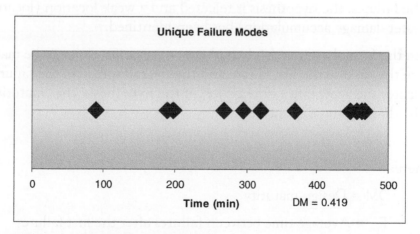

Figure 10-3: Hypothesis rejected.

The test is set up with the level one stresses set at service conditions. If the hypothesis is correct, that the product is accumulating stress damage throughout the product in a uniform way, then at level one the rate of stress damage will be uniform. Level ten of the test is set up with each stress source raise to a destruct limit or a change in the physics of failure. For example, the maximum temperature would not be raised above the glass transition point of a plastic part, and the voltage would not be raised beyond the electrical breakdown limit of key components. The destruct limit of each stress is defined as the stress level that will cause failure in only a few cycles (less than 1 hour of exposure) without changing the physics of the failure. Because the stresses at level ten are all set to destroy at a short period of time, the rate of stress damage is uniform (one life of damage is accumulated in less than 1 hour of exposure).

If the hypothesis of the test is correct (that uniform stress damage accumulation occurs in the product under service conditions), and the tenth level is set with all stress sources causing failure in a short period of time, then the rate of damage accumulation should remain uniform from level one through level ten. If a failure mechanism is accumulating damage faster than the rest of the design at or near service conditions, then that failure mechanism will exhibit the failure well before the rest of the design fails. In other words, if a failure occurs earlier than the rest of the failures, the hypothesis is rejected and a weak location (location of faster damage accumulation) has been identified.

From the formulation of this hypothesis, a quantification can be made. Since the time to the first failure and the overall spread of the failures indicates the acceptance or rejection of the hypothesis, the "maturity" of the design can be quantified as[4]:

$$DM = T_{ave}/T_{min}$$

Where:

DM = Design maturity

T_{ave} = Average time between failures after the first failure

T_{min} = Minimum time-to-failure

---

[4] Entela, "Design Maturity Algorithm," U.S. Patent: 6,247,366.

$$T_{ave} = ((T2 - T_{min}) + (T3 - T2) + (T4 - T3)...(T_{max} - T_n)) / (count - 1)$$

$$T_{ave} = (T_{min} - T_{max}) / (count - 1)$$

Where:

T_{max} = Maximum time-to-failure

T_x = Time-to-failure of failure number x

Count = Count of failures

(See Figure 10-4.)

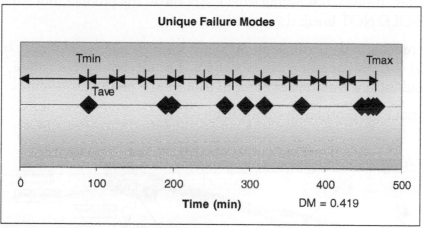

Unique Failure Modes

Tmin

Tave

Tmax

0   100   200   300   400   500

Time (min)   DM = 0.419

Figure 10-4: Failure mode progression.

Another way to view this is that DM is the average potential for improvement by fixing one failure. DM therefore provides a means of quantifying how well the product met the hypothesis.

However, DM only tells part of the story. The maturity of the design provides a measure of how much better the product could get under the accelerated stress conditions. A relative measure of a product's life is also needed if products are going to be compared. This is the Technological Limit (TL) and can be defined by removing failure modes and recalculating the DM until the DM is less then a target value. The time of the first remaining failure mode is the technological limit. We'll discuss more about technological limit later.

More About Stress

Keep in mind that the term STRESS is being used here in a more general way. Stress is considered to be more than just load over area. It is anything that can damage the product. That understanding leads to a logical question to ask when approaching the task of identifying stress sources for an FMVT.

What can break the product?

Notice a couple of things about this question. It does not ask, "What do we expect the product to see?" or, "What did I design the product to handle?" Instead, it is asking what CAN break the product, not what SHOULD NOT break the product.

Referencing and coordinating the mechanism of failure column (Chapter 5) with this question can be helpful. The mechanisms of failure should include the stresses that can damage the product.

Consider the toothbrush from Chapter 5.

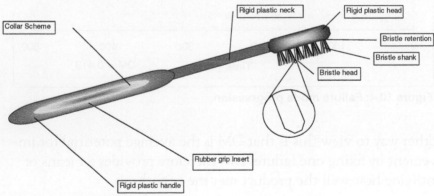

Figure 10-5: Toothbrush design

In the DFMEA example that was started there, the following mechanisms of failure were noted:

- Impact
- Thermal cycle
- Chemical attack/material incompatibility

- Fatigue
- Sharp radius

Now a couple of notes: First, not everything in the list should be included in the stress source list. Only stresses that are external to the design should be included. This is because the test is designed to determine what failures exist in the design. In this case, the "Sharp radius" mechanism of failure would not be included.

The second thing to note is that this list is not complete. Consider some of the other things that can damage a toothbrush:

- Radiant heat (sun lamp)
- Biting
- Dry toothpaste buildup at base of bristles
- Humidity
- Immersion (water)
- Gripping
- Fire/open flame
- Abrasion

Now consider how general some of the mechanisms of failure from the DFMEA are, and make them more specific.

Impact

    Dropping

    Caught in door close

    Object dropped on

Thermal cycle

    Hot

    Cold

    Ramp rate

    Boiling

Chemical attack/material incompatibility

> Toothpaste
>
> Mouthwash/prep
>
> Hydrogen peroxide
>
> Bleach

Fatigue

> Bending (head – left/right, up/down)
>
> Torsion

Now take all of them and begin to make a table.

Table 10-1:

Stress		Service Conditions	Maximum Level (Destruct)	Application Method
Dropping	m	1	100	height above floor
Caught in door close	N	1	10	force applied in scissor fashion with two edges of wood
Object dropped on	kg	2	20	blunt object from 1 m
Hot	degC	50	150	chamber
Cold	deg C	4	-60	chamber
Ramp rate	degC/min	2	20	chamber
Boiling	min	5	50	heated water
Toothpaste	ml	0.5	5	Measured and mixed with water and scrubbed in
Mouthwash prep	ml	10	100	measure and spray on
Hydrogen peroxide	ml	10	100	measure and spray on
Bleach	ml	10	100	measure and spray on
Bending (head – left/right, up/down)	deg	5	45	air cylinder and rubber grip on head
Torsion	N-m	2	20	clamp and air cylinder
Radiant heat (sun lamp)	delta Deg C	20	70	Heat lamp controlled to the given surface temperature rise
Biting	N	1	10	sharp ceramic nubs clamped around head
Dry toothpaste buildup at base of bristles	ml	0.5	5	Measured, applied and allowed to dry
Humidity	rh	5	95	chamber
Immersion (water)	min	1	10	complete submersion
Gripping	N	1	10	clamp and air cylinder on handle
Fire / Open flame	deg C	n/a	n/a	Not a service condition stress - DROP
Abrasion - stroke	1/min	60	600	surface with simulated teeth bumps moved back and forth in across bristles
Abrasion - pressure	N	1	10	surface with simulated teeth bumps moved back and forth in across bristles

In filling in the table, any ridiculous sources of stress are dropped. In case of fire, the toothbrush's service conditions do not include fire at all. So why list fire in the first place? In brainstorming, you want people to think past the normal "expected stresses," and get to the nonintuitive stresses. The easiest way to increase the likelihood of capturing all relevant stresses that can break the product is to brainstorm until you begin to get ridiculous stresses.

Table 10-2:

Stress	Units	Service Conditions	Maximum Level (Destruct)	Application Method
Dropping	m	1	100	Plate hanging to contact tooth brush in vibration
Caught in door close	N	1	10	force applied in scissor fashion with two edges of wood - once per level between levels (manually)
Object dropped on	kg	2	20	blunt object contacting tooth brush during vibration
Hot	deg C	50	150	chamber
Cold	deg C	4	-60	chamber
Ramp rate	degC/min	2	20	chamber
Boiling	min	5	50	heated water - boil tooth brush between levels
Tooth paste	ml	0.5	5	Measured and mixed with water and scrubbed in between levels
Mouth wash/prep	ml	10	100	measure and spray on
Hydrogen peroxide	ml	10	100	measure and spray on
Bleach	ml	10	100	measure and spray on
Bending (head–left/right, up/down)	deg	5	45	Mass attached to head to induce motion from vibration machine
Torsion	N-m	n/a	n/a	combined with bending above
Radiant heat(sun lamp)	delta Deg C	20	70	Heat lamp controlled to the given surface temperature rise
Biting	N	n/a	n/a	sharp ceramic nubs clamped around head - combined with mass in bending above
Dry toothpaste buildup at base of bristles	ml	n/a	n/a	combined with Tooth Paste above
Humidity	rh	5	95	chamber
Immersion (water)	min	n/a	n/a	complete submersion - not needed coverd in boiling
Gripping	N	1	10	clamp and air cylinder on handle
Fire/ Open flame	deg C	n/a	n/a	Not a service condition stress - DROP
Abrasion - stroke	1/min	n/a	n/a	add simulated teeth to plate above - strocking from vibration machine
Abrasion - pressure	N	n/a	n/a	add simulated teeth to plate above - strocking from vibration machine

In one project I worked on, the team identified a variety of stresses including bugs (the six-legged kind). There was some debate about what the service levels of bugs were, but it was finally decided that they do tend to nest in warm places (like electronic boxes) and that the exoskeletons and other debris could cause electrical problems. A bait shop was the source for purchasing wax worms and crickets that were euthanized in a nitrogen environment, blended (not a fun job) and a "nest" was built in the logical nooks and crannies of the electrical enclosure. As it turned out, not only were bugs detrimental to the operation of the device, but they also produced some electrical current as they decayed—literally, they became a semi-conducting battery/capacitor.

The point is, the brain storming needs to get past the expected and into the slightly bizarre. Use the service conditions to determine what stresses to drop. When in doubt, include the stress source.

The next thing to do is to examine the stresses for the destruct limit. How high would a stress go before: 1) The part would break, or 2) The physical limit of the stress is reached. For example, the door closing force necessary to destroy the product is easily determined by taking one toothbrush and closing the door harder and harder until the handle breaks. Physical limits of a stress can be reached in some cases like the toothpaste; there is a logical amount of toothpaste that can be placed on a toothbrush, after which any more toothpaste would fall off. In this case, the toothpaste will either break the toothbrush (destruct limit), or it never will.

Finally the stresses should be examined for the application method. Notice that the application methods listed in the first table are for each individual stress. Often there are stresses that can be combined when all stresses are applied together. For example the impact force from dropping the toothbrush and the abrasion force could be combined.

One final note about stress: There are some times when zero stress is more damaging than a higher level. For example, fretting corrosion on a low voltage contact can be caused because of vibration abrading away the plating on the contact followed by an absence of vibration. The vibration can actually keep the potential corrosion from building up to the point where contact is lost.

More About Failures

Since the goal of an FMVT is to identify failure modes, the definition and instrumentation of failure modes is critical. The DFMEA and the effects of failure are a good place to look for ideas of what affects to instrument for.

More About Setup and Execution

Setting up an FMVT requires preparing the fixturing and instrumentation for all of the stress sources and failure modes identified. Don't worry, you'll do it wrong the first time. It is often a good idea to drop some of the more difficult stress sources and instrumentation for the first attempt and keep it simple. Caution: dropping a stress has been proven to change the failure modes found.

The accompanying CD-ROM contains a couple of virtual examples of FMVT executions.

Note that many FMVT's are conducted on a single part. We will address how we handle the potential for outliers in the data analysis section.

More on Data Analysis

The results of an FMVT start with the incident log, including a description of the incident, the time at which the incident is observed, and the test level of the incident.

All incidents should be recorded regardless of their perceived relevance at the time! Why? Experience has shown that an incident early in the test that does not appear to be relevant will become a critical clue later on when a failure manifests itself.

See Table 10-3 for an example of the incident log from the toothbrush.

Several things can immediately be determined from this data. The first plot to look at is the failure mode progression. For convenience, it is of-

Table 10-3: Incident log for the toothbrush example.

Event	Description	Time Under Test	Level	Failure Number
1	Bristles A falls out	20	1	1
2	Bristles B falls out	100	2	1
3	Rubber grip insert delaminates at	120	3	2
4	Bristles F falls out	150	3	1
5	Rubber grip insert delaminates at	160	3	2
6	Bristles E falls out	180	4	1
7	Rigid plastic head splits	190	4	3
8	Rubber grip insert falls out	240	5	2
9	Bristles D falls out	260	5	1
10	Bristles C falls out	290	5	1
11	Rigid plastic head splits to neck	300	6	3
12	Rigid plastic neck comes off of handle	340	6	4
13	Handle cracks along nit line	355	6	5

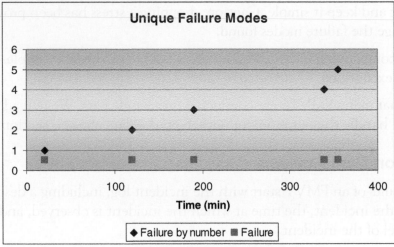

Figure 10-6: Failure mode progression of toothbrush.

ten wise to plot the data two ways, vs. failure number and linearly. The failure mode progression plot shows the relative timing of the FIRST occurrence of the different failures. Notice that the first failure of the toothbrush occurs very early. The next two failures are somewhat close together, and the last two are clustered further out. We will formally analyze this distribution later, but in general, you can see that the first failure is very critical (level 1 of an FMVT is service conditions).

In addition to the plot, the Design Maturity (DM) and technological limit can be calculated.

Table 10-4: Design Maturity (DM) and Predicted Design Maturity (PDMx).

DM	4.1875
PDM1	0.652778
PDM2	0.434211
PDM3	0.044118

You can see that the DM of 4.1875 (meaning that fixing one failure would give an average of over 400% improvement in the life of the product) reinforces the observation from the plot that the first failure is critical. However, you can also see that fixing the next two failures would also have (on average) a beneficial impact: over 60% and 40%, respectively, for PDM1 and PDM2 (Predicted Design Maturity from fixing the first x failures). However, by the time the first three failures are fixed (PDM3 = 0.044118), the potential for improvement is very low (less than 5%). This can be seen in the graph by looking at the last two failures. Fixing the second to last would improve the part very nominally.

So far, we have looked only at the first occurrence of the failure. The next item to look at is the repetition of the failures. The histogram of the failures vs. the level in which the failures occurred vs. how often the failure occurred provides several important clues. With a statistical test, outliers are identified by their deviation from the mean relative to the standard deviation of the population. During the FMVT (which is often on only one system), when a failure occurs, the item that fails is repaired or replaced. Naturally, if the failure that occurred is inherent to the design and not just an artifact of the particular fabrication, the failure will occur again and again. In the histogram of the toothbrush failures, you can see that the bristles falling out occurs over and over again. However, the rubber insert delaminating does not occur as often.

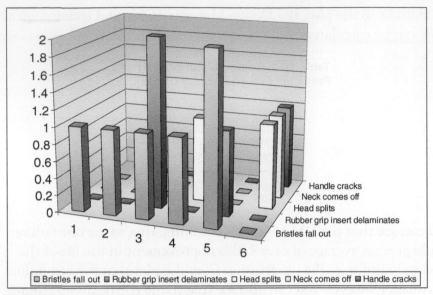

Figure 10-7: Histogram of toothbrush failures.

In some cases, the histogram will show a failure that occurs once very early, and then never repeats. This type of failure is usually the result of a fabrication issue. This does not mean the failure should be ignored. In fact, the differentiation between repeatable, design inherent failures and fabrication issues is one of the very powerful results of an FMVT. Knowing that a particular fabrication step can easily result in an early failure allows the production process to target controls on critical steps. The histogram from an FMVT is an effective tool for sorting out between design inherent failures and fabrication issues.

In the case of the toothbrush, the failure mode progression is failure small. A more complex failure mode progression may be the controller data shown in Figure 10-8.

In this case, the design maturity calculations become very important in sorting out the failures. (See Table 10-5.)

You can see here that the design maturity and the predicted design maturity is a bit more complex. You can also see that fixing the first failure

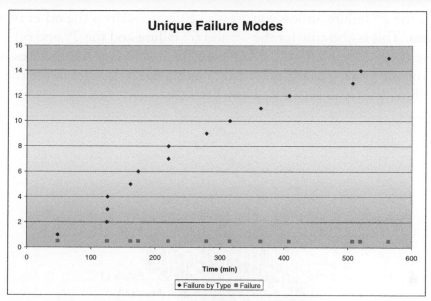

Figure 10-8: Controller failure mode progression.

Table 10-5: Controller design maturity and predicted design maturity.

	DM	Change
DM	0.767857	
PDM1	0.270154	0.497703
PDM2	0.289683	-0.019529
PDM3		
PDM4	0.248148	0.067869
PDM5		
PDM6	0.194005	0.055038
PDM7	0.221719	-0.027715
PDM8	0.170251	0.051469
PDM9	0.158095	0.012156
PDM10	0.13843	0.019665
PDM11	0.127451	0.010979
PDM12	0.056213	0.071238

reduces the potential for improvement from around 76% to 27%, but then fixing failures makes the measure *worse*. This is because the second three failures are clustered together. This is important in evaluating what failures to fix and which ones to leave. There is not much sense in fixing the 2nd failure, unless you are also going to address the other two failures. This is also true for the 5th and 6th failure and the 7th and 8th failure (keep in mind that PDM5 means the first 5 have been addressed and you are looking at the potential of fixing the 6th). After fixing the 7th failure, the trend is continuously better.

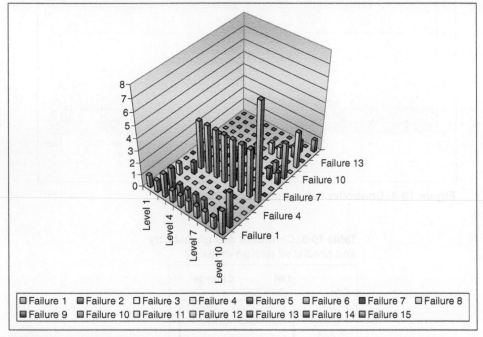

Figure 10-9: Histogram of controller failures.

The histogram for the controller data shows more clearly the use of the histogram to sort failures. Notice that the first two failures are very repeatable and happen early—definite targets for improving the design. The 3rd–5th failures all occur once very early and then never repeat. These failures are likely fabrication errors. Knowing this changes the decisions that may be made about the 2nd failure. From the failure mode progression and the PDM analysis above it was noted that the 2nd–4th failures should be addressed as a group. However, the histogram indicates that the 3rd and 4th failures are fabrication related. They will be addressed through production controls separately from the 2nd failure, which is a design inherent failure mode.

One other consideration can be made in examining failures from a complex system, and that is to sort the failure mode progression based on severity or subsystem. In this case, the controller failure mode progression is separated into three progressions, for mild, medium and severe failures. This can also help identify which failures are worth fixing and which are not.

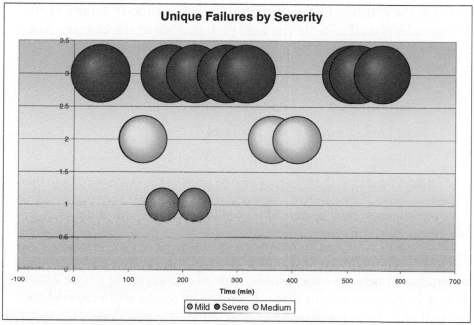

Unique Failures by Severity

Time (min)

Mild Severe Medium

Figure 10-10: Ranked failure mode progression of controller.

Comparison FMVT

FMVT measures the progression of failures and estimates, which failures should be fixed, and which are normal end of life wear out of an optimized design. By comparing the failure mode progression of an existing design and a new design, a comparison of the quality of the new design to the old design can be made on two key points: time to first failure and optimization or failure mode progression.

Provided that known field references, such as: Best in Class, Best in World or Best in Company designs are available, a simple FMVT comparison can be conducted in two ways.

Method One: Time to First Failure

In this method, a design exists for which relative field performance is known (for example, a 3-year warranty rate of 0.96%), an FMVT is conducted on a sample of the existing design. After the test is complete, the test level at which the first relevant failure occurred is identified. A new sample of the existing design and a sample of the new design are tested at the identified level until both fail. The time-to-failure of the two samples is used to scale the field performance of the existing design to provide an expected field performance of the new design.

For example:

> Existing design MTTF in the field is 67 months.
>
> First hard failure occurred at level 7.
>
> Existing design time-to-failure at level 7   243 minutes
>
> New design time-to-failure at level 7   367 minutes
>
> Estimated new design MTTF = (67 months / 243 minutes) 367 minutes = 101 months

This method can be used with multiple reference samples to establish an expected range. In other words, instead of comparing to one existing design, compare to two or three designs. The results will vary slightly but will provide an estimated range in which the new design is expected to perform.

Method Two: Failure Mode Progression Comparison

In this method, the existing design and the new design both go through a complete FMVT and the comparison is conducted on the full list of failure modes and their relative distribution. This is a more thorough comparison because it evaluates the time to first failure and also the efficiency of the design. An estimate on the improvement in the design for field performance can be made from the design maturity quantification of failure mode progression.

FMVT Life Prediction – Equivalent Wear and Cycle Counting

When using a broad range of stress sources in an accelerated test plan, and a broad range of failure mechanisms is present, a life prediction can be made IF the damage accumulated is proportional for all of the different mechanisms.

An example of this is a test conducted on an automotive window regulator, in which metal-on-metal wear, metal fatigue, plastic-on-metal wear, and threaded fastener torque loss were the failure modes. In this particular case, the accumulated damage on these different failure mechanisms could be documented and compared to the damage accumulated during a controlled life condition test. It was determined that 8 hours in an FMVT correlated to the same damage accumulated in a 417 hours life test, which correlated to one life in the field.

Using this technique, a life test was designed that took only 8 hours AND quantified the life of the product through the equivalent wear.

When a door or a closure of some type is present, processing the time domain data of the door motion can make a life prediction. The motion of the closure is analyzed to "count" equivalent cycles. In this method, a cycle from the field conditions of a product (for example, a toaster oven door) is analyzed for its characteristics in displacement, velocity, acceleration, voltage and current. A profile of one cycle is then determined as a set of conditionals (a door "open", acceleration going negative, then positive, displacement passing through a given point and so forth). These conditionals are then used to count the number of times a cycle ("open" event) is caused during a random (vibration, voltage and cycling) fatigue test. The cycles are then plotted as a histogram against their severity (number of cycles with a 4 G peak spike). In this way, Minor's rule can be applied to the equivalent number of cycles and the amount of equivalent life can be determined. This method has been used on products to get a quantified life cycling test from 200 hours down to less than 8 hours.[5]

---

[5] "Life Estimating Techniques for Failure Mode Identification Testing Methods," SAE Congress: 2002-01-1174.

FMVT Warranty

FMVT can be used for troubleshooting warranty issues. To accomplish this, a standard FMVT is run on the product. The emphasis is put on identifying and applying all possible stress sources. Once the standard FMVT is conducted, two possibilities exist. Either the warranty issue was reproduced, in which case the troubleshooting can go to the next stage. Otherwise, a significant fact has been established. The warranty issue is due to a stress source that was not identified or applied. If this is the case, then the additional stress source(s) must be identified and the FMVT rerun.

Once the warranty issue has been reproduced in the full FMVT, then a narrower test of limited stresses and levels is determined that will reproduce the warranty problem in a short period of time. Usually, a test that produces the warranty failure mode on the current design in only a few hours can be produced. This test can then be used to test design solutions. Once a design solution is identified, a full FMVT should be conducted.

More on Vibration

Failure mode verification testing requires the application of a wide range of stress sources to a product. Stress sources are sources of damage to the product. The stress sources are applied to the product to induce failure modes that are inherent to the design of the product but would not otherwise be easily detectable. Stress sources are typically vibration, temperature, voltage, pressure, chemical attack and so forth. Of all of these stress sources, the most difficult and the most critical is vibration.

Vibration is present in the working environment of any product, from automobiles, to airplanes, from desktop computers to soap dispensers. In addition to being prevalent, vibration is inherently destructive. Even low levels of vibration can cause significant damage to a product. Vibration is able to do this for three reasons:

1) Vibration is a repeated event that occurs as little as several times a second to as much as tens of thousands of times a second.

Imagine a debt that with payments made at the rate of $.01 per second (1 cent every second). At that rate, a $10,000.00 debt would be paid off in 10,000,000 seconds or just over 115 days. Vibration works the same way, doing very little (1 cent is not much) but doing it over and over again very fast.

2) Vibration is significant because of the natural frequencies that are inherent in every product. A natural frequency is the way a part "rings" like a tuning fork or a fine crystal wine glass. When a tuning fork is struck it rings. The sound it makes is produced by the motion of the forks. This motion is called the *mode shape*; the shape a part moves naturally in when stimulated. Mode shapes are extremely important for vibration damage to a product. To break a product from fatigue, the product must be strained, like bending a metal coat hanger back and forth until it breaks. The tuning fork's mode shape is the shape the product will bend the most in and bend easiest in. If vibration is applied to the product at its natural frequency, the bending that results in the product will be significantly higher than if the vibration was applied at some other frequency. This relationship is expressed in the following graph:

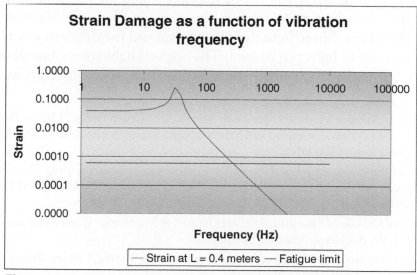

Figure 10-11: Potential strain damage as a function of vibration.

Notice that in this example, applying vibration below the natural frequency (98 Hz) has a significant effect on the level of strain damage. Applying vibration above the natural frequency results in very little strain damage. Applying vibration directly at the natural frequency causes a much larger rate of strain damage.[6] [7] [8] [9] [10]

The importance of the vibration spectrum can be summed up this way: vibration that is at or just below the natural frequency of the product being tested will provide significant contributions to the accumulated strain damage in the product.

3) Vibration is critical because it contributes to the accumulation of damage from other stress sources. Vibration exacerbates thermal stresses, bearing surface wear, connector corrosion, electrical arcing and so forth.

Vibration is critical to properly stressing a product, but vibration is made up of several components: amplitude, spectrum, and crest factor to name a few. As seen previously, the frequency at which the vibration is present is important to how effective it is in producing damage in a product. In reality, vibration exists at multiple frequencies simultaneously, not at one frequency. These multiple simultaneous frequencies are called the *vibration spectrum*. In this sense, mechanical vibration is much like light. Shine light through a prism and the different colors (frequencies) of light can be seen. The original light source has all of the frequencies present simultaneously. Vibration spectrums vary as

[6] "Effect of Random Vibration Spectra on Test Product Components," Gilbert Bastien, IEEE Components, Packaging and Manufacturing Technology Society 2000, Accelerated Testing Workshop.

[7] "Evaluating Damage Potential Producing Vibration Environments using The Shock Response Spectrum," George Henderson, IEEE Components, Packaging and Manufacturing Technology Society 2000, Accelerated Testing Workshop.

[8] "Tutorial On Use of Spectrum Analyzers for 6DOF Machines," George Henderson, IEEE 2002, Accelerated Testing Symposium.

[9] "Generating Random Vibration for Accelerated Stress Testing," Wayne Tustin, IEEE 2002, Accelerated Testing Symposium.

[10] J.P. Holman, *Experimental Methods for Engineers*, (McGraw-Hill, 1984).

widely as the colors of lights. Choosing the right spectrum requires understanding the natural frequencies of the products being tested (so that most of the spectrum is at or below the natural frequencies of the part) and knowing the spectrums that the vibration equipment can produce.

All vibration equipment has a spectrum from zero to infinity. An office table has a vibration spectrum from zero to infinity. The question is how much AMPLITUDE exists at the different frequencies. There are three physical limits that govern the amplitude on all vibration equipment: displacement, velocity and acceleration. See Figures 9-8 through 9-10.

Displacement

At low frequencies, the displacement is the limiting factor on acceleration amplitude. The laws of physics dictate that a given frequency and acceleration requires a certain displacement. At 1 Hz (1 cycle per second), a 4 g's acceleration (acceleration 4 times greater then earth's gravity) would require nearly 1 meter (just over 1 yard) of displacement. Most vibration equipment claims a spectrum down to 5 Hz. At 5 Hz, a 4 g acceleration requires 6.2 cm (2.44 inches) of displacement. Evaluating the low frequency capabilities of a machine is then easy. What is its maximum possible displacement?

Velocity

Velocity is a limit for some machines in the mid-frequency range. Servo-hydraulic machines are limited in the maximum velocity of the pistons used to drive the machines because of the maximum flow rate of the hydraulic supply.

Acceleration

Acceleration is a limit on all machines based on two factors: The mass that is being moved, and how strong the components of the machine are. Force is equal to mass times acceleration. The maximum force the machine can produce (directly or kinetically) will limit the maximum acceleration. The other consideration is that the machine must be able to withstand the forces necessary to move the mass. Maximum accelerations are usually advertised for a machine.

The amplitude at higher frequencies is also governed by damping and control methods. With most servo-hydraulic multi-axis vibration tables, the limiting factor is the natural frequency of the vibration equipment. Operating at the natural frequency of the vibration equipment would be very damaging to the capital investment. For this reason, most multi-axis servo hydraulic equipment is limited to 70 Hz, while some have limits up to 350 Hz. Air hammer tables use the natural frequencies of the table itself to reach very high natural frequencies; the repeatability and uniformity of the tables are subject to the natural frequencies of the particular table. The upper-end of the spectrum on the FMVT machine is limited by hysteresis in the vibration mechanism dampening out the spectrum produced by the mechanical recursive equations used to produce the vibration.

Reliability and Design Maturity

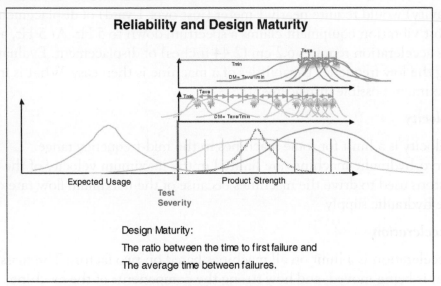

Figure 10-12: The relationship between reliability and design maturity.

Design maturity as a measure was first developed to address the issue of objectively sorting failure modes, especially in a contract situation between companies. However, the historical measure for contracts has been the statistical reliability measure discussed in earlier chapters. Design Maturity and statistical reliability are related.

On the accompanying CD-ROM is a PowerPoint slide show of a discussion between reliability and design maturity. The bottom line is this: statistical reliability and design maturity measure two orthogonal characteristics of the same whole.

In other words, the individual failures that would be seen as a failure mode progression (along the x-axis) relate to the failures that should not be seen in a fully censored test (see Chapter 6) through their respective accelerated reliability curves (see Chapter 8). With FMVT, the conscious decision is made to find the failure modes and rank them relative to each other and their stress levels and to not know the time-to-failure in the field or their acceleration curves.

The confidence in the failure mode progression comes from knowing that each individual failure does have an acceleration curve (stress vs. time-to-failure), and that curve limits how early the failure can occur in the FMVT and still meet the service time requirements. A failure that occurs at level 2 would require an impossibly steep curve to meet life requirements under service conditions.

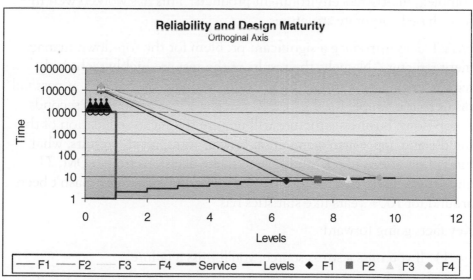

Figure 10-13: Reliability and design maturity exist on orthogonal axis.

Business Considerations

FMVT provides meaningful data (information) for the design engineer, especially during the development stages of a product. The comparison testing, equivalent wear and cycle counting provide a means for the tool to be used for reliability estimating. For warranty chases, FMVT has proven to be a very useful tool. The design maturity analysis and the graphical tools help provide a means of relating the data to context and decisions that must be made (see Key fact #1 from Chapter 1).

The entrepreneurial business (innovative part instead of commodity) will find this tool to be especially helpful when the technology used to fabricate it is new, and understanding of how to test is limited. The brainstorming session combined with an abbreviated DFMEA has jump-started the level of understanding on many new technology validation programs.

Commodity businesses have found great use in the comparison application of FMVT to conduct design of experiments extremely quickly on complex, multistress environment products. This has worked well in team-based corporate structures.

FMVT does introduce a significant problem for the top-down management structure. Namely, the results of the test are highly technical, require engineering thought and evaluation and are not a clear pass/fail result that is simple for the 10,000 ft. view VP to grasp in 15 seconds. A top-down business structure will require significant education of the middle and upper management on how to interpret the results, what kind of questions to ask, and how to make decisions from a DM, TL and failure mode progression. It's not rocket science, but it hasn't been around for 100+ years like statistics has.

Key facts going forward:

1) If you miss a stress source, you'll miss a failure.

2) If you're not going to spend some time understanding the results, don't do the test.

11

Computer and Math Modeling

As far as the laws of mathematics refer to reality, they are not certain; and as far as they are certain, they do not refer to reality.
—Albert Einstein

Computer and math models fall into a couple of simple categories:

1) Corollaries of physical laws
2) Approximations

Math Models

Laws of motion, momentum, velocity, displacement and acceleration are very clearly defined math models. Most high school students conduct various simple experiments exploring the laws of physics that Sr. Isaac Newton discovered. Now we know that his laws of motion only apply at speeds much slower than the speed of light, and Einstein's general relativity takes over as speeds increase. It will not at all surprise me if in my lifetime physicists find the "unified field theory" that explains all forces in the universe from one simple force. The derivations of math models from these laws of physics are often fairly accurate and can be solved in a deterministic way.

For example, the laws of motion lead to:

$$v = a * t + vo$$
$$d = v * t + do$$

where:

v = velocity

vo = initial velocity (at t = 0)

a = acceleration

t = time

d = displacement

do = initial displacement (at t = 0)

These equations are fairly simple to solve and can be used to determine the actions of an object that is not under the influence of other forces.

Physical properties of materials can also lead to very deterministic calculations. The deflection on a simple beam can be found using equations derived from material properties and integration along the beam.[1] These can be used to provide simple equations for particular situations.

$v = (Px^2 / (6EI))(3L - x)$

$v' = (Px / (2EI))(2L - x)$

P = Point force

v = deflection in the y direction at x

v' = slope in at x

E = Elastic modulus

I = Area moment of inertia

L = Length of beam

(See Figure 11-1.)

Unfortunately, most products in real life are more complicated than the previous two examples. Using more involved models that account for complex behavior in the bending and strain of a general geometry can approximate the behavior of the product, but not duplicate it exactly.

---

[1] James Gere and Stephen Timoshenko, *Mechanics of Materials*, (Wadsworth Publishing Company, 1984).

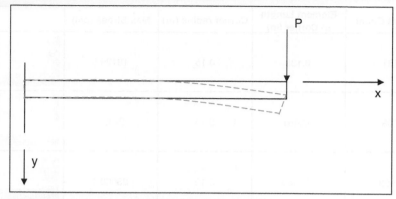

Figure 11-1: Simple cantilever beam in bending.

Finite Element Analysis (FEA)

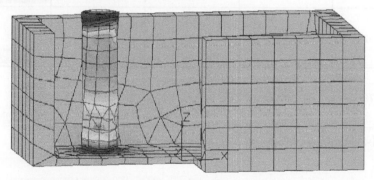

Figure 11-2: Computer model of an enclosure with a plastic boss for a self-tapping screw. The elongation in the boss has been exaggerated.

For example, a computer model of the stress on a complex plastic part as it is loaded in multiple directions is found through finite element analysis. This model has 1,679 solid elements defined by 1,050 nodes. The stress matrix used to solve the problem is a matrix of 2,760 equations and unknowns. This yields a linear approximation of the stress field between each node. This would be a very simple model requiring only a few seconds to run. A larger assembly can easily exceed hundreds of thousands of elements and equations. The accompanying CD-ROM contains the decoder output from the model below. The CD-ROM also has more examples of models and some animations of models responding to stress.

Element Count	Element Length in Corner (in)	Corner radius (in)	Max Stress (psi)	
231	0.135	0.15	18135.5	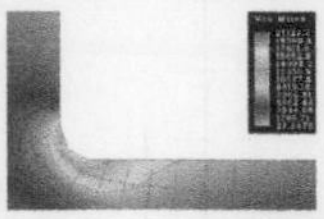
709	0.0461	0.15	22127	
1403	0.040681	0.15	23060.1	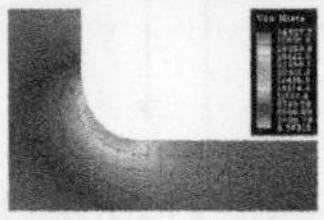
5822	0.0203	0.15	23541.9	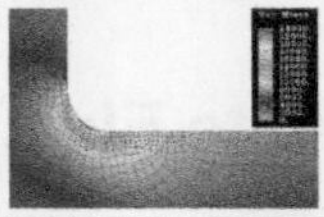
15657	0.01389	0.14	24519.6	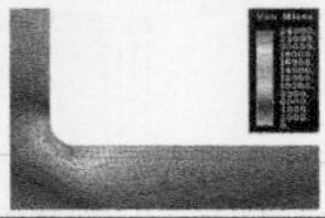
17778	0.010171	0.15	25418.3	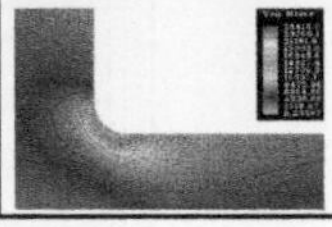

Figure 11-3: Model accuracy vs. element size and count.

The accuracy of the computer model can be checked by refining the model. Figure 11-3 shows the progression of a model with a corner using a progressively higher number of elements.

By running a model at several element resolutions, a computer model can be verified to be approaching an approximate solution. (See Figure 11-4.)

Computer modeling can be used to determine stress, strain, deflection, mass, moment of inertia, vibration mode shapes, fluid flow, temperature distributions, electrical and magnetic characteristics and more.

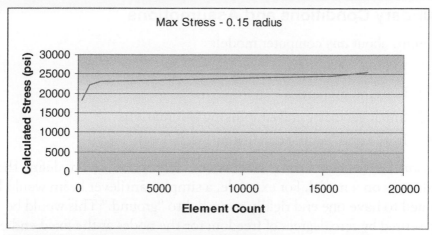

Figure 11-4: Element count vs. calculated stress.

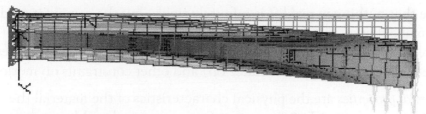

Figure 11-5: Finite element model of a simple cantilever beam.

```
1**** CONTROL INFORMATION

            number of node points           (NUMNP)   =          1050
            number of element types         (NELTYP)  =             1
            number of load cases            (LL)      =             5
            number of frequencies           (NF)      =             0
            geometric stiffness flag        (GEOSTF)  =             0
            analysis type code              (NDYN)    =             0
            solution mode                   (MODEX)   =             0
            equations per block             (KEQB)    =             0
            weight and c.g. flag            (IWTCG)   =             0
            bandwidth minimization flag     (MINBND)  =             0
            gravitational constant          (GRAV)    =     3.8640E+02

            bandwidth minimization specified

1**** NODAL DATA

  NODE      BOUNDARY CONDITION CODES        NODAL POINT COORDINATES
  NO.    DX  DY  DZ  RX  RY  RZ      X           Y           Z           T
  ----------------------------------  ----------  ----------  ----------  ----------

     1    1   1   1   1   1   1   1.371E-01  -1.325E+00   0.000E+00   0.000E+00
     2    1   1   1   1   1   1   4.919E-01  -1.325E+00   0.000E+00   0.000E+00
     3    1   1   1   1   1   1   8.467E-01  -1.325E+00   0.000E+00   0.000E+00
     4    1   1   1   1   1   1   1.201E+00  -1.325E+00   0.000E+00   0.000E+00
     5    1   1   1   1   1   1   1.556E+00  -1.325E+00   0.000E+00   0.000E+00
     6    1   1   1   1   1   1   1.911E+00  -1.325E+00   0.000E+00   0.000E+00
  ........
```

Table 11-1: Output header from a finite element analysis decoder. The full file is on the accompanying CD-ROM.

Boundary Conditions and Assumptions

Key items about *any* computer model:

1) Boundary conditions
2) Material properties
3) Small geometry and simplifying assumptions
4) Initial and loading conditions

Boundary conditions are the mathematical definitions used to define the constraints on a model. For example, a simple cantilever beam would be assumed to have one end rigidly mounted to "ground." This would be represented by zero degrees of freedom for the nodes at the fixed end of the beam.

Note that nothing in real life is infinitely rigid—but the computer model can make that assumption. Boundary conditions can include all six degrees of freedom (three translations, three rotations), contact surfaces (like a tabletop for a ball to bounce on) and other constraints on motion.

Material properties are the physical characteristics of the material the model is simulating. This is a source for many mistakes. Most published material data is insufficient for a complete model of a material's behavior. In addition, most plastics will behave differently in tension and compression and depending on how they are molded. It is wise wherever possible to use material properties gained from the as-molded material in a geometry similar to the intended part.

Watch out for material properties that are different between compression and tension or are different depending on strain rates. Most general published material properties are based on a "static" or slow strain rate. For example, the standard reference for rigid plastics is ASTM D638, which defines the modulus of elasticity based on the "linear" region of the elongation curve of the plastic under constant strain. The modulus is one of the most important material properties for computer modeling. However, most plastics experience a different elastic modulus for compression vs. tension. ASMT D790 is a test for flex modulus which gives an elastic modulus based on the assumption that the tension and compression modulus is the same. When reviewing the material's properties for plastics, comparing the tensile modulus (from ASTM D638) and the flex modulus (from ASTM D790), will indicate how different the compression and tension modulus are. Unfortunately, for many FEA packages, this is data and not information because they can only use one modulus for both compression and tension.

Table 11-2: Physical properties for ABS

Subcategory: ABS Polymer; Polymer; Thermoplastic

Key Words: Poly(Acrylonitrile Butadiene Styrene)

Material Notes:
Information provided by Ebbtide Polymers.

No vendors are listed for this material. Please click here if you are a supplier and would like information on how to add your listing to this material.

Physical Properties Metric English Comments

Density 1.04 g/cc 0.0376 lb/in³ ASTM D792
Linear Mold Shrinkage 0.006 cm/cm 0.006 in/in MD; ASTM D955
Melt Flow 1.2 g/10 min 1.2 g/10 min ASTM D1238

Mechanical Properties

Hardness, Rockwell R 102 102 ASTM D785
Tensile Strength @ Break 40 MPa 5800 psi ASTM D638
Tensile Strength @ Yield 40 MPa 5800 psi ASTM D638
Elongation at Break 25 % 25 % ASTM D638
Elongation at Yield 3 % 3 % ASTM D638
Tensile Modulus 1.9 GPa 275.5 ksi ASTM D638
Flexural Modulus 2.6 GPa 377 ksi ASTM D790
Flexural Strength 63 MPa 9140 psi ASTM D790
Izod Impact, Notched 4.38 J/cm 8.2 ft-lb/in ASTM D256

Small geometry and simplifying assumptions can cause some unintended consequences. Modeling very small geometry (like a screw boss in a large plastic enclosure) can be very difficult. Automatic modeling software and computer modeling technicians may "simplify" this geometry in order to make the model run better in the computer. The problem comes if the geometry simplification is in a key area of interest.

Another example is corners. The computer model can appear to have an infinitely sharp corner.

Element Count	Element Length in Corner (in)	Corner radius (in)	Max Stress-infinite radius (psi)	
16081	0.015555	infinite	27800.2	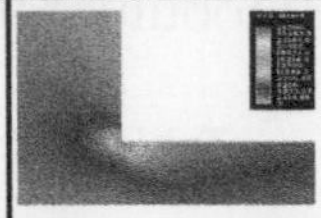
14866	0.034569	infinite	26009.4	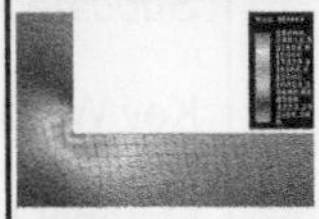
5002	0.0622	infinite	21096.1	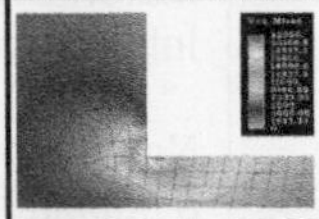
2612	0.07778	infinite	19077.8	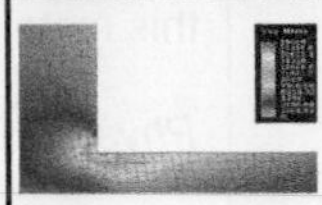
518	0.1555	infinite	14964	
82	0.6222	infinite	9706	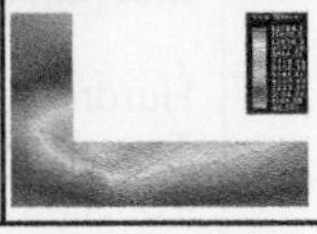

Figure 11-6: Corner sharpness. A computer model can make a corner infinitely sharp. However, the element size creates a mathematical equivalent to the round so that large elements around the corner can have a similar effect to a truly rounded corner.

In fact, the corner as modeled will have a radius equivalent to half the element length. Of course, as the elements are made smaller and smaller, the stress at the corner will approach infinity—a perfectly sharp corner is impossible in real life.

Initial and loading conditions are extremely important to the performance of the model. Just as the boundary conditions can distort the results of a computer model, so can the initial conditions or the way loads are simulated. For example, a 10 N load applied to a button could be simulated by a point load at a node. The effect would be a force of 10 N spread across ¼ of each of the surrounding elements—for a finite area and a resulting pressure. Or, the 10 N load could be applied across 4 nodes, resulting in a larger area.

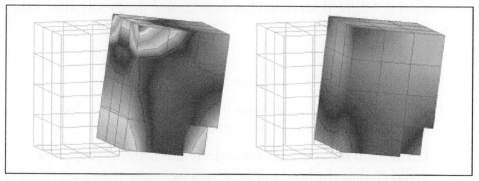

Figure 11-7: Point loading vs. distributed load. Notice the artificial stress riser on the left-hand model.

The resulting stress from the two different conditions is very different.

All in all, computer models are very powerful tools, but they can be very precisely wrong. One of the big downfalls of computer models is that they can always make very cool looking pictures and animations (just look on the CD-ROM). But the wow factor does not make good science. Always ask the following questions about the computer model:

1) What is the hypothesis or question that is being addressed by the model?

2) What is the uncertainty of the results relative to the question or the hypothesis?

3) How would changing the boundary conditions, material properties or other assumptions affect the results?

4) Would the change in results due to reasonable changes in the boundary conditions change the conclusion or decisions that are made from the information?

For example, if a computer model was used to evaluate the potential compliance of a dashboard to Federal Motor Vehicle Safety Standards (FMVSS) head impact requirements and the nominal material properties for the foam in the dash were used, the computer model could very well show a better performance than if the "as-molded" foam properties were used in the model. In addition, the rigidity with which the computer model holds the dash (boundary conditions) drastically affects the impact results.

Dash:

1) Hypothesis: FMVSS compliance.

2) Uncertainty is due to changes in foam characteristics from molding and in–dash mounting stiffness.

3) Changes in the boundary conditions and material properties could change the head impact forces by _____.

4) The computer results will have to meet FMVSS compliance even with the reasonable changes to be taken with some confidence.

In this example, the hypothesis and sources of uncertainty are clear (items 1 and 2). Item 3 will have to be determined through multiple runs of the computer model. Of course if the nominal run fails, there is not much need to go further. Once the multiple runs are completed, the hypothesis can be evaluated against all of the results to come to one of three conclusions: The hypothesis is rejected; the hypothesis is rejected for some cases; the hypothesis is accepted for all cases. In the case of the second possibility (the hypothesis is rejected for some cases), some physical testing will need to be accomplished to determine which model is appropriate. Note that some laws and common sense (not the same thing) will require physical testing even if the computer model says it will work.

Business Considerations

In the past, computer modeling was the domain of rich companies and large budgets, rooms full of expensive Unix machines and geeks in the dark. Today, computer-modeling software can be purchased relatively inexpensively, and computer-modeling services through the internet compete globally. This makes computer modeling accessible for most any size business and for most business models. Both commodities and innovative products can benefit from the tool, but for different reasons. In the case of commodities, computer-modeling software can be used to drive reduction in material content, number of components or improved ease of assembly. For the innovative design, it is often the computer modeling tools that make it possible for an individual to work through the invention process and develop new ideas.

The one problem for all of the users of computer modeling programs is avoiding the wow factor. In fact, all of the tools described in Chapters 5–11 have one significant limitation: they can't do the whole job on their own. Computer models need basic material data input, preferably from the as-molded part. They should also be checked after the fact for accuracy against a physical test. A physical test that is able to impose stress and stimuli to the product that are not easily modeled in the computer can expand the usefulness of the computer results. Chapter 12 will explore the possibilities of hybrid tests that combine two or more accelerated testing techniques to leverage the data.

12

Hybrid Testing

If you limit your choices only to what seems possible or reasonable, you disconnect yourself from what you truly want, and all that is left is a compromise.

—Robert Fritz

Often test methods are viewed as individual items to be executed in a vacuum—a single hypothesis tested using a single method. It is very reminiscent of Newtonian physics. A mass in motion remains in motion. We have an image in our minds of a mass (size, shape, color doesn't matter) in a complete void, moving. The mass keeps moving because there is nothing to interact with it at all. Here's an interesting question: If the mass in Newton's Law remains in motion because there is nothing to interact with it, then what is the motion measured against?

With tests, we often conduct them in the naive void. The reality is a lot closer to Einstein's theories of general relativity; everything is perceived relative to the observer. In other words, real life interacts in complex ways. This does not diminish from the absolutes in the universe, just our perceptions of them. We often think of relativity and Heisenberg's uncertainty principle as getting in the way of the goals of our testing—to discover the truth. In this chapter, we will examine what happens if we take advantage of the complex interactions in order to learn more.

If you look at Table 12-1, you will see that there are at least 21 pairs of tests based on the seven tools discussed in this book. There could be 28 if you assumed that you could pair a DFMEA with another DFMEA and

Table 12-1: Possible combinations of hybrid accelerated tests.

Group		DFMEA	FC	SS	AR	HALT	FMVT	CM
	DFMEA		1	1	1	1	1	1
	FC			1	1	1	1	1
	SS				1	1	1	1
	AR					1	1	1
	HALT						1	1
	FMVT							1
	CM							
DFMEA	DFMEA							
	FC			1	1	1	1	1
	SS				1	1	1	1
	AR					1	1	1
	HALT						1	1
	FMVT							1
	CM							
FC	DFMEA		1	1	1	1	1	1
	FC							
	SS				1	1	1	1
	AR					1	1	1
	HALT						1	1
	FMVT							1
	CM							
SS	DFMEA		1	1	1	1	1	1
	FC			1	1	1	1	1
	SS							
	AR					1	1	1
	HALT						1	1
	FMVT							1
	CM							
AR	DFMEA		1	1	1	1	1	1
	FC			1	1	1	1	1
	SS				1	1	1	1
	AR							
	HALT						1	1
	FMVT							1
	CM							
HALT	DFMEA		1	1	1	1	1	1
	FC			1	1	1	1	1
	SS				1	1	1	1
	AR					1	1	1
	HALT							
	FMVT							1
	CM							
FMVT	DFMEA		1	1	1	1	1	1
	FC			1	1	1	1	1
	SS				1	1	1	1
	AR					1	1	1
	HALT						1	1
	FMVT							
	CM							
CM	DFMEA		1	1	1	1	1	1
	FC			1	1	1	1	1
	SS				1	1	1	1
	AR					1	1	1
	HALT						1	1
	FMVT							1
	CM							

21	Pairs
126	Tripples
147	Total

DFMEA	Design Failure Mode & Effects Analysis
FC	Fully Censored Statistical Reliability
SS	Step Stress
AR	Accelerated Reliability
HALT	Highly Accelerated Life Testing
FMVT	Failure Mode Verification Testing
CM	Computer Modeling

somehow leverage more information out of it. I won't say you can't (in fact I know you can and should), but this book has to have some limits and some material for a second edition. The 21 pairs of tests would be using the two tests to leverage the information from each test to address a wider range of hypotheses or questions than the tests could accomplish on their own. There are also 126 triples; three tests leveraged against each other to expand the information gained.

There is more than one way to combine any two or three tests.

We will examine some of the possible interactions, but these are only examples. Knowing that the possible interactions exist, and having a means of defining them, will make it possible to recognize circumstances when new combinations will be useful.

Process Diagrams: To keep the interactions down to a shorthand notation, we will first examine each testing tool as a process block having inputs, noise factors, controls and outputs.

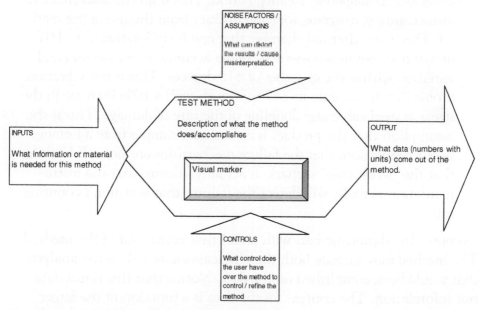

Figure 12-1: General process diagram for a test.

Test Method: The test method is one of the methods that has been discussed from Chapters 4–11. A brief description of what the test method does is included. A "visual marker" is included but not necessary. The visual marker is a plot or diagram that captures some salient point about the method and provides a quick visual reference for what the P-diagram (process diagram) is. This will be especially useful in Chapters 13 and 14 when many P-diagrams will be combined with decision points and full validation plans developed. (Besides, I'm a visual person and I find pictures an easier quick reference.)

Inputs: The stuff you need to have and know to accomplish the test method. Information and product or material is usually sufficient.

Controls: The methods, apparatus, strategies, analysis and practices that the method provides to manipulate the test method, mitigate noise factors and optimize the timing and output of the test method.

Noise Factors/Assumptions: Reality, period. This is all the stuff that can distort, annoy, frustrate, fool and distract from the use of the method. This is the thermal chamber that can be programmed to $1/10^{th}$ of a degree, but uses a thermal couple accurate to ±2 degrees and variation within the chamber of ±10 degrees. This is the vibration profile that is accurate to ±10% even though a 10% increase in the stress in steel can cause doubling of the rate of damage. This is the assumption that the product is an accurate sample from a homogeneous population, that the failure mechanisms are not self-healing, that the "worst case" scenario is actually known, that the instrumentation actually will detect the failures that you aren't counting on.

Outputs: The data (numbers with units) that comes out of the method. The method may include both the physical test as well as the analysis that would be accomplished on the data. Notice that this is just data, not information. The context for the data is a function of the larger project, not simply a result of the test itself. The data (numbers with units) becomes information within the context of the project when the behavior of sentient beings change.

Fully Coupled and Partially Coupled Hybrid Tests

Given the inputs, noise factors and outputs of each test method, you have probably anticipated how the discussion of hybrid tests can proceed. Every test has limitations, sensitivity to noise factors and limited output. So use the strengths of one or more methods to support or mitigate the weaknesses of another. It's the quid pro quo of test methods. Figure 12-2 shows a hybrid test involving HALT and FMVT.

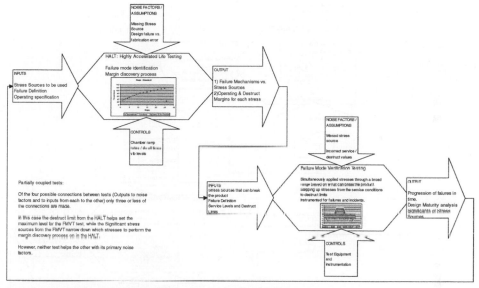

Figure 12-2: Hybrid test involving HALT and FMVT.

Examine the relationship between Highly Accelerated Life Testing (HALT) and Failure Mode Verification Testing (FMVT). This can be called a *partially coupled testing pair*. The output of one supports the input of the other in a limited way. In this case, one of the key inputs for an FMVT test is the destruct limit of the stress source. If the destruct limit is set properly, the FMVT results will be much more accurate. The destruct limit is a primary output from the HALT testing. Likewise, one of the limitations of the HALT is the time it would take to run a large number of different stress sources. The FMVT can run a large number of stress sources and establish their relevance. With key stress sources established, the HALT can be run to determine the destruct limit on the limited number of stresses for which the data (destruct limit) would be useful information.

Notice that under this scenario, a full test program may involve running a brief HALT to establish destruct limits for certain stress sources, then a full FMVT to determine the full failure mode progression, then a follow-up HALT to establish operational and destruct limits for key stresses identified in the FMVT. This would be a *HALT–FMVT–HALT grouping*.

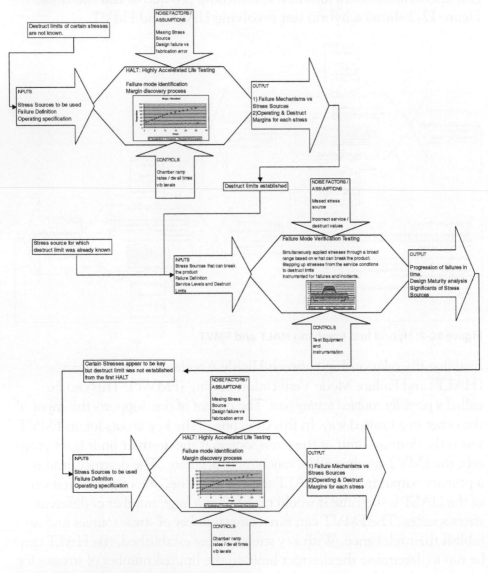

Figure 12-3: HALT–FMVT–HALT hybrid test.

One more note about the P-diagram (all of which are available on the accompanying CD-ROM): The inputs, controls, noise factors and so forth should be refined for each application. The P-diagrams presented here are general. Chapter 14 will provide some more specific examples.

The Field as a Test Method

One test method that is not often considered is the *field*. After all, the ultimate test is to put the full population of product into service in the field for multiple lives and document what happens. Obviously this has some drawbacks, but it is often the case that current production product will have a new version in development. Using the field data as an input into another test is logical.

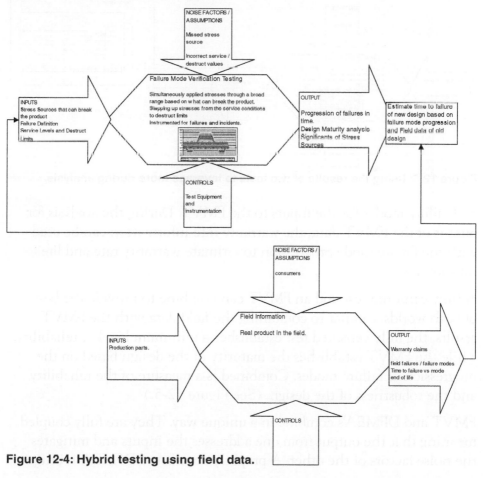

Figure 12-4: Hybrid testing using field data.

In this case, the field data along with the FMVT provides a partially coupled testing pair. The warranty claim information, field failure modes and failure rates vs. time can help establish the stress sources

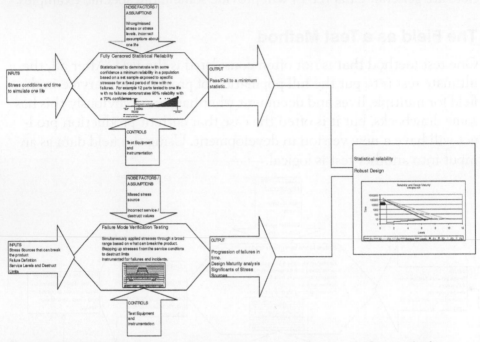

Figure 12-5: Using the results of two tests to leverage more during analysis.

and failure modes for the inputs to the FMVT. During the analysis for output of the FMVT data, the warranty rate information can be used with the failure mode progression to estimate warranty rate and life information.

A fully censored test and an FMVT can combine to provide the best of both worlds. Similar to combining the field data with the FMVT results, the fully censored test establishes a minimum level of reliability, while the FMVT establishes the maturity of the design based on the progression of failure modes. Combined is a measure of the reliability and the robustness of the design. (See Figure 12-5.)

FMVT and DFMEA's combine in a unique way. They are fully coupled, meaning that the output from one addresses the inputs and mitigates the noise factors of the other. A properly conducted DFMEA is unique

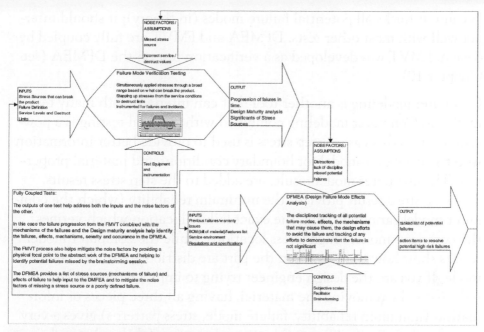

Figure 12-6: Fully coupled test. The inputs and outputs fully complement the inputs and assumptions of the other test.

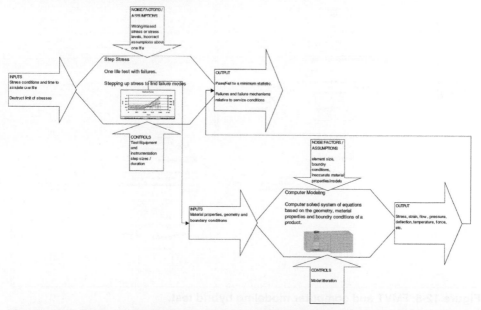

Figure 12-7: Step stress and computer model hybrid test. The step stress results influence the computer model boundary conditions, and the results of the computer model influence the analysis of the step stress results.

because it tracks all potential failure modes (in theory); it should inter-act well with most other tests. DFMEA and FMVT are fully coupled by design. FMVT was developed as a verification test on the DFMEA (see Chapter 10).

Computer modeling is another tool that can interface with many test methods. Computer modeling combined with physical testing is a pow-erful tool. In this case, a step stress is used to provide better information into the computer model for boundary conditions and material proper-ties. The computer model results are added to the step stress results. The step stress result provides the minimum reliability (which the com-puter model cannot do) and the exact failure location which verifies the computer model's high stress locations, while the computer model results show how the stresses in the part are distributed for the failure mode. If you are the design engineer trying to improve the product, or reduce cost by removing the material, having all three pieces of infor-mation (minimum reliability, failure mode, stress patterns) gives a very complete picture of what to address or what can safely be changed.

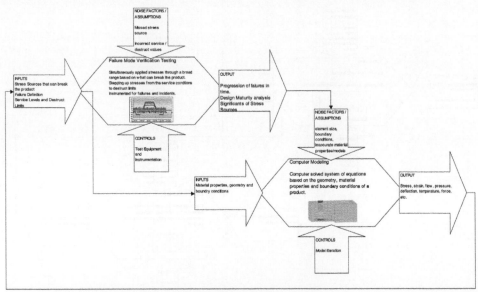

Figure 12-8: FMVT and computer modeling hybrid test.

Computer modeling and FMVT are often used in conjunction. Computer modeling has a difficult time with a large number of stress sources and complex loading and simplifying assumptions are often made. FMVT is well suited to large numbers of stress sources and complex interactions. The inputs to the FMVT (stress sources, potential failure modes) immediately lead to the fixturing and instrumentation; this can be used to define the boundary conditions for the computer model. Likewise, the results of preliminary computer models can be used to help guide the instrumentation. For example, a computer model can predict the location of maximum deflection due to load or vibration. That location can be used for instrumentation during the FMVT. The out put of the FMVT helps control some of the noise factors for the computer model by verifying the locations of failures, performance of geometries (deflection, resonance, strain) and checks the complex interactions not captured in the computer model.

Key facts going forward:

1) Do not limit your choice of test methods to what seems possible or reasonable based on one method. You will disconnect yourself from what you truly want, and all that is left is a compromise. Define what information you want and define combinations of test methods that will leverage their respective strengths, mitigate their weaknesses and offer up solid information (data in the context of your project that will affect your behavior).

2) Combining methods is not limited to just input and output. Look for interactions in the midst of the test plan and ways to mitigate the limitations.

3) Do not limit your thinking to physical test methods. Consider all information generation tools.

13

Validation Synthesis Model

Analysis and synthesis ordinarily clarify matters for us about as much as taking a Swiss watch apart and dumping its wheels, springs, hands, threads, pivots, screws and gears into a layman's hands for reassembling, clarifies a watch to a layman.

—Author Unknown

Imagination is the beginning of creation. You imagine what you desire, you will what you imagine and at last you create what you will.

—George Bernard Shaw

As discussed back in Chapter 2, the wealth of test methods, procedures, guides and examples makes choosing a test for a particular application difficult and overwhelming. The task of designing a validation plan that provides the information needed to make sound engineering and business decisions from research through to production with confidence seems an impossible task. This chapter will discuss how to use the key information, business structure, product type, supply chain position and timing to structure a validation plan that supports the quality and engineering development plan (the business model). A framework will be established for translating the key information parameters and timing needs from the quality control plan into validation plan goals and the means of achieving them. In other words, how to make the pile of gears and gizmos in Chapters 4–11 work together for something approximating a rational purpose.

Purpose: End or aim; design; intention; matter in question. *–vt.* And *i.* (purposing, purposed). To propose; to intend.[1]

> *Many persons have a wrong idea of what constitutes true*
> *happiness. It is not attained through self-gratification but*
> *through fidelity to a worthy purpose.*
>
> —Helen Keller

Is important to remember when setting out to devise a plan, a purpose, or intention to drive first for the greatest ideal that can be imagined. You will fall well short of the goal, so imagine where you will be if you fall short of your aim at mediocrity?

The primary purpose of engineering is to turn to practical use the science and technology in an upright and just manner in the service of humanity.[2]

The primary purpose of testing is to produce information so sound decisions can be made in the pursuit of engineering and the business of using engineering to bring a product to market.

In setting out to establish a validation plan, it is advisable to start from the information needed to fulfill the oath (or obligation) of the engineer. What information is needed to ensure that the research, feasibility, development, testing, production, use and disposal of the product will be based on sound science and technology and will be just and upright in the service of humanity? Answer this question and you are well on the way to a good validation and testing plan.

I realized in writing the above paragraphs that there are some who will be offended, or consider this to be idealized nonsense. I recognize this. Early in 2003, I chaired an SAE workshop on "Lean Validation"[3] which discussed many of the topics covered in this book and focused on the problem of implementing accelerated testing within the supply chain of

---

[1] *New Webster's Dictionary*, (Paradise Press, Inc., 2001)

[2] Obligation of the Engineer, Order of the Engineer.

[3] "CL001: Lean Validation Engineering Clinic Bridging the Gap—Moving Toward Lean Validation," SAE 2003.

the automotive industry. In the discussion, many people expressed frustration that they could not change the way things were being done in practice because they were too low in the line of authority and had to just do what they were told and not rock the boat. I pointed out that attitude is in direct contradiction to the oath (obligations) of the engineer.

There have been a few times (thankfully, only a few) when I have had to make a fundamental choice between speaking up or keeping quiet for fear of my job. I have always taken the choice of speaking up; I have never lost a job, promotion or raise for doing this. In fact, just the opposite. I will observe that standing up for what is right is not easy and requires tact, and humility—the kind that acknowledges that I might be the one who is wrong. But I have never regretted standing up for what I believed to be right, even if I felt threatened.

Insisting on driving towards the ideal is not only the right thing to do, but I have found it leads to better engineering, more robust business practice, a greater satisfaction in work done, and greater respect from people you work with. I know that I fall short of the ideals I set for myself, and I know that there may be people who read these two paragraphs and utter the word "hypocrite" because of some failing I have committed that they may remember. Well, duh! Of course I fail and fall short. That is why it is important for me to strive toward the ideal and not some lesser goal. For those of you out there who are perfect, you can aim for mediocrity and hit it. I have to aim for the ideal.

OK, off the soapbox. What was that question?

The Primary Question

What information is needed to ensure that the research, feasibility, development, testing, production, use and disposal of the product will be based on sound science and technology and will be just and upright in the service of humanity?

The most logical place to go for this list is the Design Failure Mode Effects Analysis (DFMEA) or some other quality control and tracking document that tracks *all* of the potential issues, including safety, regulations, potential failure modes and open issues. In this case, we will use the DFMEA. From the DFMEA, there are a couple of things to look at. First, consider whether the DFMEA is written well. A well-written DFMEA that is started at the beginning of a project (when it is supposed to be) will need an action item for each of the potential failure modes. Keep in mind that all of the product requirements should show up as potential failures in the DFMEA with NOT added (the light bulb filament provides continuity between the poles: NOT). In addition, all safety and regulation requirements should be captured the same way.

The DFMEA is also the place to capture the product type (commodity vs. innovation), the business model, the place in the supply chain and any other critical assumptions.

Commodity assumptions in the DFMEA: With commodities, the key to business success is the reliability vs. cost. The product must be reliable with low cost. To that end, the DFMEA should reflect key hypotheses expressed as potential failures: the product costs more than \$1.25 per piece; the product experiences more than 1:10,0000 first-year failure rate.

Business model assumption in the DFMEA: Examine the business model and the key assumptions to make it work. If product name recognition and customer loyalty are critical, then the DFMEA should reflect potential failures such as: Product logo is not legible at end of life; product soils consumer during replacement; product is not uniquely recognizable at 100 yards. Obviously, these types of failures would have nothing to do with the usability of the product by the consumer, but would impact on the business model—in this case, if the business model is based on customer loyalty and brand recognition.

Supply chain position assumptions in the DFMEA: A tier one supplier must achieve certain objectives with a product being supplied to a particular OEM. OEM-directed requirements must be reflected in the DFMEA: product molded in color does not match painted metal surface of other automotive interior trim; product tracking number not traceable through OEM tracking system.

In each case, it is critical that the key assumptions be embodied in the DFMEA, so that the information flowing from the DFMEA into the hypothesis matrix and into the final test plan reflects the information needed to validate the assumptions made in the business and engineering plan.

Table 13-1: Hypothesis matrix.

Potential Failure	Mechanism of Failure						
	Vibration	High Temperature	Cold Temperature	Temperature Ramp	Humidity	Salt air	Torque
Filament loss of continuity	Vibration will not cause Filament loss of continuity	High Temperature will not cause Filament loss of continuity	Cold Temperature will not cause Filament loss of continuity	Temperature Ramp will not cause Filament loss of continuity	Humidity will not cause Filament loss of continuity	Salt air will not cause Filament loss of continuity	Torque will not cause Filament loss of continuity
Corrosion	Vibration will not cause Corrosion	High Temperature will not cause Corrosion	Cold Temperature will not cause Corrosion	Temperature Ramp will not cause Corrosion	Humidity will not cause Corrosion	Salt air will not cause Corrosion	Torque will not cause Corrosion
Seal loss between glass and base	Vibration will not cause Seal loss between glass and base	High Temperature will not cause Seal loss between glass and base	Cold Temperature will not cause Seal loss between glass and base	Temperature Ramp will not cause Seal loss between glass and base	Humidity will not cause Seal loss between glass and base	Salt air will not cause Seal loss between glass and base	Torque will not cause Seal loss between glass and base
Base Thread does not match receptacle	Vibration will not cause Base Thread does not match receptacle	High Temperature will not cause Base Thread does not match receptacle	Cold Temperature will not cause Base Thread does not match receptacle	Temperature Ramp will not cause Base Thread does not match receptacle	Humidity will not cause Base Thread does not match receptacle	Salt air will not cause Base Thread does not match receptacle	Torque will not cause Base Thread does not match receptacle
Base di-electric allows current flow	Vibration will not cause Base di-electric allows current flow	High Temperature will not cause Base di-electric allows current flow	Cold Temperature will not cause Base di-electric allows current flow	Temperature Ramp will not cause Base di-electric allows current flow	Humidity will not cause Base di-electric allows current flow	Salt air will not cause Base di-electric allows current flow	Torque will not cause Base di-electric allows current flow
Electrolytic reaction with receptacle material	Vibration will not cause Electrolytic reaction with receptacle material	High Temperature will not cause Electrolytic reaction with receptacle material	Cold Temperature will not cause Electrolytic reaction with receptacle material	Temperature Ramp will not cause Electrolytic reaction with receptacle material	Humidity will not cause Electrolytic reaction with receptacle material	Salt air will not cause Electrolytic reaction with receptacle material	Torque will not cause Electrolitic reaction with receptacle material
Streaking in glass coating / un even lighting	Vibration will not cause Streaking in glass coating / un even lighting	High Temperature will not cause Streaking in glass coating / un even lighting	Cold Temperature will not cause Streaking in glass coating / un even lighting	Temperature Ramp will not cause Streaking in glass coating / un even lighting	Humidity will not cause Streaking in glass coating / un even lighting	Salt air will not cause Streaking in glass coating / un even lighting	Torque will not cause Streaking in glass coating / un even lighting
Loss of print on bulb	Vibration will not cause Loss of print on bulb	High Temperature will not cause Loss of print on bulb	Cold Temperature will not cause Loss of print on bulb	Temperature Ramp will not cause Loss of print on bulb	Humidity will not cause Loss of print on bulb	Salt air will not cause Loss of print on bulb	Torque will not cause Loss of print on bulb

From each potential failure mode, there are several potential mechanisms of failure. It is the potential mechanisms of failure that should be addressed (vibration causes the light bulb filament to lose continuity between the poles). Each potential mechanism of failure should be formed into a hypothesis H: vibration will not cause the filament to lose continuity between the poles of the light bulb.

Keep in mind that each potential failure mode has several potential mechanisms of failure. Therefore, a table structure is logical with each potential failure listed down the side, and the mechanisms of failure listed across the top. This is called the *hypothesis matrix*. Notice that each potential failure can be combined with each mechanism to form a hypothesis. Even the mechanisms that cannot reasonably produce the given failure will make a logical hypothesis (torque will not cause filament loss of continuity).

Match each hypothesis to the tests that could fulfill the hypothesis in the table for each potential failure mode and each failure mechanism; there will often be several tests. Include the accelerated tools from this book as appropriate as well as material tests, performance test and company specifications that seem reasonable. Some of the cells in the table will be marked as N/A because the failure cannot reasonably be caused by the mechanism. When a cell can be filled in, list all of the test methods that can be used to test the hypothesis.[4]

For each potential test for a given hypothesis, determine three things:

1) Relative cost of conducting the test at each level of development.

2) Relative cost of NOT knowing that the hypothesis is false until each level of development.

3) The confidence with which the particular test supports the hypothesis.

OK, so what's that mean? Well, look at Chapter 3 for Key facts #1 and #2, and look at Chapter 5 for Key fact #3.

---

[4] Porter, A., "What Process For What Part?" SAE 2000 World Congress, March 2000, Detroit, MI, USA. Session: Accelerated Testing Conference (Part A&B).

The cost of conducting a test at each level should be fairly straightforward. The cost of a fully-functional prototype during the research phase is astronomical compared to the cost of a production intent part during production ramp. However, the cost of a production intent material sample has a relatively low cost at any stage.

The cost of NOT knowing that the dissimilar materials will not work goes up with each level of development and becomes astronomical by the time production ramp and production starts.

The confidence that a particular test supports the hypothesis is based on how well (or how thoroughly) the test checks the hypothesis against the final design in the field. For example, testing a material sample that has been coated with the production intent protective coating for electrolytic reaction during the research phase is a relatively low cost test. But the test does not account for the effects of geometry, residual stress, abrasion during service and so forth. Therefore, it is an inexpensive test, that (if passed) supports the hypothesis that the material will not cause a dielectric reaction very early in development (when the cost of NOT knowing is low), but it does so with low confidence (low detectability in the DFMEA detection column). (See Figure 13-1.)

It may be that some simple pass/fail compatibility or performance tests should be conducted early when they are inexpensive to mitigate risk and then confirmed later on more complete, representative product when the cost for the larger test is lower (but finding the hypothesis is false would be expensive).

Undoubtedly, it is a large task to go through every potential failure mode, determine the list of tests, determine the cost, risk and detectability for each test, balance cost and risk with inexpensive early tests combined with more complete complex tests. So we need some tools to reduce the size of the problem.

Timing

One of the biggest drivers for any development plan in modern times is time-to-market. Applying a timeline to the development cycle will quickly eliminate test methods that simply cannot fit the timing. Examine the time it takes to complete the test vs. the sweet spot for the test. If the sweet spot is during design validation but your timing only allows 2 weeks from prototype availability to production ramp, then any test that cannot provide information in much less than 2 weeks (so there is time for contingencies) should be removed from the hypothesis matrix or its sweet spot should be adjusted. See Figure 13-2. Remember, if you move a test out so that the time to conduct the test matches the time available in the development level, make sure that the results of the test is still *information* and not just data.

Efficiency

Look at your hypothesis matrix with the list of potential tests. Notice that many of the hypotheses can be addressed with one test. For example, a computer model may be able to address several issues concerning deflection, heat buildup and so forth. On the other hand, there are other tests that can only address a limited number of issues: ASTM cyclic corrosion testing. The following steps will prioritize an efficient list of tests.

1) Find the one test that satisfies the greatest number of hypotheses.

   a. Circle or highlight this test in each cell where it occurs.

   b. If there is a tie, choose the test that best meets detectability and timing.

2) Find the one test that satisfies the greatest number of hypotheses in the remaining cells.

   a. Circle or highlight this test in each cell were it occurs (of those with no other highlighted test).

   b. If there is a tie, choose the test that best meets detectability and timing.

3) Repeat #2 until all cells have a highlighted test.

Now look carefully at the list of tests you have highlighted. There is a risk that the detectability from the DFMEA will not be good for some of these tests. In that case, augment the tests with one or more of the other tests (again highlight or circle the ones you use) in the cells where the detectability number from the DFMEA is low. Using this method on the light bulb in the following example reduces 39 hypotheses to three tests.

Table 13-2: Two-test risk mitigation scheme (see Figure 13-1).

	Research	Feasibility	Development/Design	Design Validation	Production Ramp	Production Validation	Production	Service
Material Sample Compatibility Test- Low Detectability								
Cost	1	1	1	1	1	1	1	1
Risk	1	1	2	3	4	5	8	10
FMVT on Production intent with all stresses that could cause electrolytic reaction - HighDetectability								
Cost	10	9	8	5	3	2	2	2
Risk	1	1	2	3	4	5	8	10

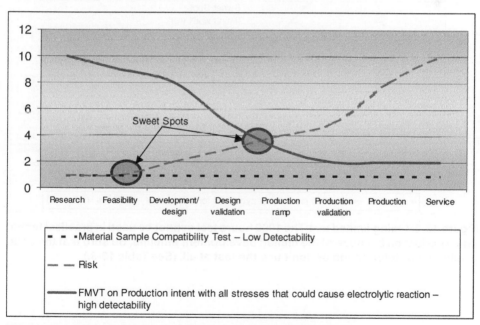

Figure 13-1: Two-test risk mitigation scheme (see Table 13-2).

Table 13-3: Timing limited test. The 1000 hours test cannot be used in practical terms until production because of the limited development timeline. Be sure that the data would still be information or don't use the test at all. (See Figure 13-2.)

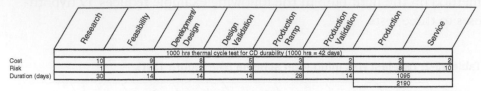

	Research	Feasibility	Development/ Design	Design Validation	Production Ramp	Production Validation	Production	Service
			1000 hrs thermal cycle test for CD durability (1000 hrs = 42 days)					
Cost	10	9	8	5	3	2	2	2
Risk	1	1	2	3	4	5	8	10
Duration (days)	30	14	14	14	28	14	1095	
							2190	

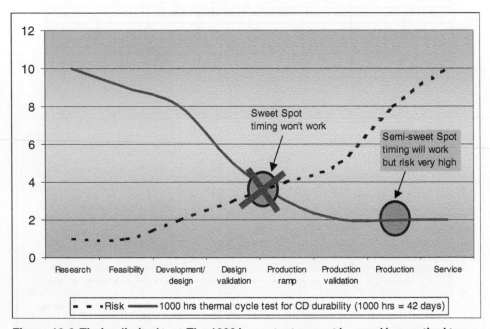

Figure 13-2: Timing limited test. The 1000 hours test cannot be used in practical terms until production because of the limited development timeline. Be sure that the data would still be information or don't use the test at all. (See Table 13-3.)

Table 13-4: Hypothesis matrix with test methods listed and optimal tests highlighted (in bold with numbers 1, 2 and 3).

Potential Failure	Mechanism of Failure						
	Vibration	High Temperature	Cold Temperature	Temperature Ramp	Humidity	Salt air	Torque
Filament loss of continuity	Vibration will not cause Filament loss of continuity	High Temperature will not cause Filament loss of continuity	Cold Temperature will not cause Filament loss of continuity	Temperature Ramp will not cause Filament loss of continuity	Humidity will not cause Filament loss of continuity	Salt air will not cause Filament loss of continuity	Torque will not cause Filament loss of continuity
	FMVT - 1 HALT ST ST CM FC Proprietary life durability test	FMVT - 1 HALT ST ST CM FC Proprietary life durability test	FMVT - 1 HALT ST ST CM FC Proprietary life durability test	FMVT - 1 HALT ST ST CM FC Proprietary life durability test	FMVT - 1 HALT ST ST CM FC Proprietary life durability test N/A	N/A	N/A
Corrosion	Vibration will not cause Corrosion	High Temperature will not cause Corrosion	Cold Temperature will not cause Corrosion	Temperature Ramp will not cause Corrosion	Humidity will not cause Corrosion	Salt air will not cause Corrosion	Torque will not cause Corrosion
	N/A	FMVT - 1 HALT ST ST FC ASTM cyclic Corrosion Proprietary life durability test	FMVT - 1 HALT ST ST FC ASTM cyclic Corrosion Proprietary life durability test	FMVT - 1 HALT ST ST FC ASTM cyclic Corrosion Proprietary life durability test	FMVT - 1 HALT ST ST FC ASTM cyclic Corrosion Proprietary life durability test	FMVT - 1 HALT ST ST FC ASTM cyclic Corrosion Proprietary life durability test N/A	N/A
Seal loss between glass and base	Vibration will not cause Seal loss between glass and base	High Temperature will not cause Seal loss between glass and base	Cold Temperature will not cause Seal loss between glass and base	Temperature Ramp will not cause Seal loss between glass and base	Humidity will not cause Seal loss between glass and base	Salt air will not cause Seal loss between glass and base	Torque will not cause Seal loss between glass and base
	FMVT - 1 HALT ST ST CM FC Proprietary life durability test	FMVT - 1 HALT ST ST CM FC Proprietary life durability test	FMVT - 1 HALT ST ST CM FC Proprietary life durability test	FMVT - 1 HALT ST ST CM FC Proprietary life durability test	FMVT - 1 HALT ST ST CM FC Proprietary life durability test	FMVT - 1 HALT ST ST FC Proprietary life durability test	FMVT - 1 HALT CM FC Proprietary life durability test
Base Thread does not match receptacle	Vibration will not cause Base Thread does not match receptacle	High Temperature will not cause Base Thread does not match receptacle	Cold Temperature will not cause Base Thread does not match receptacle	Temperature Ramp will not cause Base Thread does not match receptacle	Humidity will not cause Base Thread does not match receptacle	Salt air will not cause Base Thread does not match receptacle	Torque will not cause Base Thread does not match receptacle
Functional Test - 3	N/A	N/A	N/A	N/A	N/A	N/A	N/A
Base di-electric allows current flow	Vibration will not cause Base di-electric allows current flow	High Temperature will not cause Base di-electric allows current flow	Cold Temperature will not cause Base di-electric allows current flow	Temperature Ramp will not cause Base di-electric allows current flow	Humidity will not cause Base di-electric allows current flow	Salt air will not cause Base di-electric allows current flow	Torque will not cause Base di-electric allows current flow
	FMVT - 1 HALT ST ST FC Proprietary life durability test	FMVT - 1 HALT ST ST FC Proprietary life durability test	FMVT - 1 HALT ST ST FC Proprietary life durability test	FMVT - 1 HALT ST ST FC Proprietary life durability test	FMVT - 1 HALT ST ST FC Proprietary life durability test	FMVT - 1 HALT ST ST FC Proprietary life durability test N/A	N/A
Electrolytic reaction with receptacle material	Vibration will not cause Electrolytic reaction with receptacle material	High Temperature will not cause Electrolytic reaction with receptacle material	Cold Temperature will not cause Electrolytic reaction with receptacle material	Temperature Ramp will not cause Electrolytic reaction with receptacle material	Humidity will not cause Electrolytic reaction with receptacle material	Salt air will not cause Electrolytic reaction with receptacle material	Torque will not cause Electrolytic reaction with receptacle material
	N/A	FMVT - 1 HALT ST ST FC Proprietary life durability test	FMVT - 1 HALT ST ST FC Proprietary life durability test	FMVT - 1 HALT ST ST FC Proprietary life durability test	FMVT - 1 HALT ST ST FC Proprietary life durability test	FMVT - 1 HALT ST ST FC Proprietary life durability test N/A	N/A
Streaking in glass coating / un even lighting	Vibration will not cause Streaking in glass coating / un even lighting	High Temperature will not cause Streaking in glass coating / un even lighting	Cold Temperature will not cause Streaking in glass coating / un even lighting	Temperature Ramp will not cause Streaking in glass coating / un even lighting	Humidity will not cause Streaking in glass coating / un even lighting	Salt air will not cause Streaking in glass coating / un even lighting	Torque will not cause Streaking in glass coating / un even lighting
	FMVT - 1 HALT ST ST FC Proprietary life durability test	FMVT - 1 HALT ST ST FC Proprietary life durability test	FMVT - 1 HALT ST ST FC Proprietary life durability test	FMVT - 1 HALT ST ST FC Proprietary life durability test	FMVT - 1 HALT ST ST FC Proprietary life durability test	FMVT - 1 HALT ST ST FC Proprietary life durability test N/A	N/A
Loss of print on bulb	Vibration will not cause Loss of print on bulb	High Temperature will not cause Loss of print on bulb	Cold Temperature will not cause Loss of print on bulb	Temperature Ramp will not cause Loss of print on bulb	Humidity will not cause Loss of print on bulb	Salt air will not cause Loss of print on bulb	Torque will not cause Loss of print on bulb
	FMVT - 1 HALT ST ST FC Proprietary life durability test	FMVT - 1 HALT ST ST FC Proprietary life durability test	FMVT - 1 HALT ST ST FC Proprietary life durability test	FMVT - 1 HALT ST ST FC Proprietary life durability test	FMVT - 1 HALT ST ST FC Proprietary life durability test	FMVT - 1 HALT ST ST FC Proprietary life durability test N/A	N/A
Excessive Surface heat	Vibration will not cause Excessive Surface heat	High Temperature will not cause Excessive Surface heat	Cold Temperature will not cause Excessive Surface heat	Temperature Ramp will not cause Excessive Surface heat	Humidity will not cause Excessive Surface heat	Salt air will not cause Excessive Surface heat	Torque will not cause Excessive Surface heat
	N/A	CM FMVT - 1 HALT	CM FMVT - 1 HALT	CM FMVT - 1 HALT	FMVT - 1 HALT	FMVT - 1 HALT	N/A
Storage Temperature Causes Burst	Vibration will not cause Storage Temperature Causes Burst	High Temperature will not cause Storage Temperature Causes Burst	Cold Temperature will not cause Storage Temperature Causes Burst	Temperature Ramp will not cause Storage Temperature Causes Burst	Humidity will not cause Storage Temperature Causes Burst	Salt air will not cause Storage Temperature Causes Burst	Torque will not cause Storage Temperature Causes Burst
	N/A	CM HALT - 2 FC	CM HALT - 2 FC	CM HALT - 2 FC	CM HALT - 2 FC	CM HALT - 2 FC	N/A
Storage Pressure Causes Burst	Vibration will not cause Storage Pressure	High Temperature will not cause Storage	Cold Temperature will not cause Storage	Temperature Ramp will not cause Storage	Humidity will not cause Storage Pressure	Salt air will not cause Storage Pressure	Torque will not cause Storage Pressure
	N/A	CM HALT - 2 FC	CM HALT - 2 FC	CM HALT - 2 FC	CM HALT - 2 FC	CM HALT - 2 FC	N/A

Now let's turn this into a complete plan.

Each of the hypotheses can be listed in a Gant chart as a decision point. Lay out the timeline with each hypothesis as a decision point with an INPUT of the information needed to make the decisions. The output from the decision will be either that the hypothesis is supported or the contingency (if you included a contingency in the DFMEA) is taken. It may be useful to group the hypotheses by failure mode—but don't dumb the Gant down, the decisions (each one) should be individually made.

Now take each of the tests from the hypothesis matrix that were high-lighted or circled and add them to the Gant chart (use the P-diagrams from Chapter 12). Connect the flow of information from the start of the project through the tests and into the decisions (hypotheses), then from the decisions into the next stage of development. Be sure to line up the tests with their sweet spot and timing in the development.

It may very well be at this point that some conflicts in timing or information become obvious. Use the hypothesis matrix with its list of alternate tests to resolve any missing information and timing needs. Use the hybrid tests from Chapter 12 as a model of how to combine different tests to leverage their information.

Take one final look at the Gant chart. Two critical questions:

1) Are all of the hypotheses addressed in sufficient time for the data to be information?

2) Is the overall timing and cost of the project met?

If the answer is no, go back and look for ways to use hybrid tests to leverage data and reduce timing, add tests, or adjust the timing of the program or the level of risk you're willing to take.

To summarize the process:

1) Develop a solid DFMEA with all the critical business and engineering assumptions captured, including the business model, supply chain and commodity type influences.

   a. Be sure the DFMEA includes an exhaustive list of failure mechanisms.

b. Contingencies in DFMEAs are not standard, but will be useful when you get to the Gant chart.

2) Use the failure modes and the mechanisms to develop the hypothesis matrix.

   a. Don't worry about the hypotheses that are ridiculous (vibration will not cause an incorrect thread).

3) For each hypothesis identify a list of tests that can test the hypothesis.

   a. Include accelerated test methods, standard tests, proprietary tests and so forth.

   b. Be as exhaustive as is reasonable.

4) Rank the tests for:

   a. Detectability – how well will this test rank in the detectability criteria of the DFMEA.

   b. Cost vs. development level – how much does this test cost vs. each stage of the development cycle.

   c. Risk vs. development level – what is the risk of NOT knowing the hypothesis is wrong until the given development level.

   d. Sweet Spot – what is the sweet spot for cost vs. risk vs. development level.

5) Eliminate tests or adjust for timing.

   a. If the test cannot meet the timing requirements of the project, either eliminate the test from the hypothesis matrix or adjust its sweet spot for when in the project it could be useful.

   b. Do not keep the test if it must come so late that the results are data and not information.

6) Sort tests for efficiency.

   a. Find the one test that satisfies the greatest number of hypotheses.

      i.   Circle or highlight this test in each cell where it occurs.

      ii. If there is a tie, choose the test that best meets detectability and timing.

   b. Find the one test that satisfies the greatest number of hypthesis in the remaining cells (those with no highlighted test).

      i.   Circle or highlight this test in each cell where it occurs (of those with no other highlighted test)

      ii. If there is a tie, choose the test that best meets detectability and timing.

   c. Repeat (b) until all cells have a highlighted test.

7) Examine the tests vs. the DFMEA, and add alternative tests or combine for hybrid tests to address detectability or to mitigate risk.

8) Gant chart.

   a. Take each hypothesis and list it on the Gant chart as a decision with input and resulting decision (hypothesis supported or contingency taken).

   b. Add the tests highlighted in the hypothesis matrix to the Gant chart.

   c. Connect the flow from start of project through the tests to the decision points.

   d. Are all of the hypotheses addressed in sufficient time for the data to be information?

   e. Is the overall timing and cost of the project met?

9) Adjust the process if needed using:

   a. Alternative tests to mitigate risk/adjust timing.

   b. Hybrid tests to leverage information and reduce risk.

Downspout Generator Example

Inventions reached their limit long ago, and I see no hope for further development.

—Julius Frontinus, First Century AD.

In this chapter, we will go through a chronology of a development project from start to finish. I've picked a product that does not currently exist, but could be a real product.

Downspout Generator (DSG)

Product type: Innovation.

Business model: Entrepreneur to sell to the consumer market (the do-it-yourself crowd).

The premise of the downspout generator is to have a low cost device that can produce power from rainwater flowing through a downspout and feed it into the house through a transformer. In order for the product to be viable, it must be:

1) Easy to use – bolt the generator to the downspout and plug it in.

2) Cost efficient – the more it costs, the longer the payback period. The payback period is the cost of the unit divided by the value of the energy it produces per unit time (COST / (POWER * RATE / TIME).

3) Reliable – the flip side of #2 is that the generator can't break before it has paid for itself.

Business structure: Cross-functional team, but will supply to large national building supply chains.

Research: What would it take to make a downspout generator?

What is a downspout generator? A downspout generator is a small impeller-driven generator that hooks to the bottom of a downspout on a house or building and generates power when water flows down the downspout.

So why use a downspout generator? 1) I don't think there is such a thing, so it will be useful for the imagination. 2) The device requires some significant mechanical, material, electrical and electronic innovation. 3) If the product were real (and by the way, the only way to make this even closely feasible is for power to go upwards of $1.00 per kWh), it would be an innovation and follow the innovation business model, but would quickly become a commodity because the only way to pay for such a device is if it works flawlessly until well after its payback period.

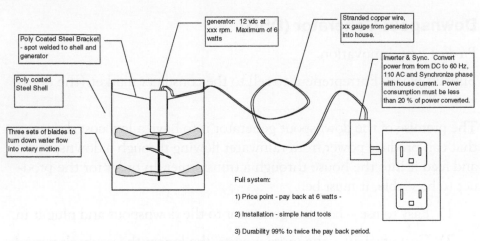

Figure 14-1: Downspout Generator (DSG) concept.

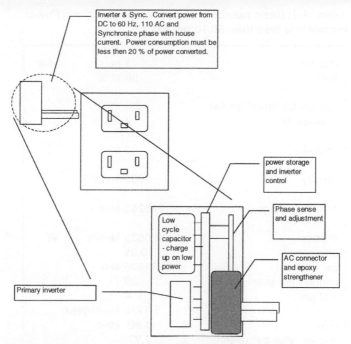

Inverter & Sync. Convert power from DC to 60 Hz, 110 AC and Synchronize phase with house current. Power consumption must be less then 20 % of power converted.

power storage and inverter control

Phase sense and adjustment

Low cycle capacitor - charge up on low power

AC connector and epoxy strengthener

Primary inverter

Figure 14-2: Exploded view of wall transformer for the DSG concept.

Basic Numbers

You can see by the numbers in Table 14-1 that the downspout generator is not practical with today's power costs. It might work on a really large building, but then the power produced would still be inconsequential. Of course if the cost of electrical power goes up over $1.00 per kWh, then the cost of making the device would probably be ten-fold more expensive also. For the sake of this illustration, we will assume that the product can be sold profitably at around $35 per unit and the power costs will be $1.50 per kWh. In that case, the product would be a neat innovation and save a little bit of money—assuming that a transformer that phases with the house current could be made very inexpensively.

Table 14-1: Basic numbers and assumptions. Note: Power currently is less than $0.10 per kWh.

assume	1 m	year
area	50 m^2	
volume for typical rainfall	2.5 m^3	
Average rate	2 hours	
potential	10 m	
density	1000 kg/m^3	
potential energy	245250 N-m	
P power	34.0625 N-m/s	W
efficiency	0.95	
per rainfall	64.71875 w-h	
rainfalls per year	20	
# of units	4	
	5.1775 kw-hr/year	
cost	$ 1.50 kw-h	
energy value per year	$ 7.77	
payback	5 years	
price	$ 38.83	

Fortunately we only care about this generator as an academic exercise involving a wide range of disciplines. So let's synthesize a testing and validation plan for this device. The first step is the DFMEA.

DFMEA

The full DFMEA is found on the CD-ROM. Here you can see a sample.

Table 14-2: Sample of the initial DFMEA for the DSG. Full DFMEA is available on the accompanying CD-ROM.

Function Item	Potential Failure Mode	Potential Effect(s) of failure	Severity	Potential Causes(s) / Mechanism(s) of Failure	Occurrence	Current Design Controls Prevention / Detection	Detectability	RPN	Recommended Action	Responsibility	Target Date
impeller											
impeller shape											
	bent	in efficient	4	leaves - debris in pipe	10	flexible carbon re-enforced polymer	6	240			
			4	installation	5	flexible carbon re-enforced polymer	6	120			
			4	thermal cycling	10	flexible carbon re-enforced polymer	6	240			
			4	water flow - fatigue	10	flexible carbon re-enforced polymer	6	240			
	broken	loss of power to shaft	8	leaves - debris in pipe	10	flexible carbon re-enforced polymer	6	480			
			8	installation	5	flexible carbon re-enforced polymer	6	240			
			8	thermal cycling	10	flexible carbon re-enforced polymer	6	480			
			8	water flow - fatigue	10	flexible carbon re-enforced polymer	6	480			
	wrong shape	in efficient	4	poor foil choice for non-laminar intermittent flow	5	Fluid flow computer model analysis software used to choose between competing shapes	6	120			
	Cost to high	Pay back period to high	8	material/fab cost to high	10	flexible carbon re-enforced polymer	6	480			

The failures in the DFMEA include particular business model related failures; for example, "costs too high" is a failure because the unit has to hit a price point for the payback period to work. The DFMEA includes contingencies. This is critical in establishing risk during validation plan synthesis.

The DFMEA delineates 28 functional parts of the DSG, and identifies 230 ways to produce 83 types of failure modes. In other words, this is a fairly small DFMEA. Notice that the PREVENTIVE method and the CONTINGENCY is filled in, but not the DETECTION method. This is because the validation plan is being synthesized in this chapter—the validation plan will provide the detection methods.

With the DFMEA done, we can begin creating the initial hypothesis matrix. On the CD-ROM, the DFMEA, the initial hypothesis matrix, test hypothesis matrix and so on are each on successive tabs of the DSG project spreadsheet. (See Table 14-3.)

Notice that we now have 83 potential failures and 26 mechanisms of failure (some of the DFMEA mechanisms were dropped because they are not physical tests, but design flaws that would be found during testing). For example, one mechanism of failure was "poor foil choice for nonlaminar intermittent flow" for the failure of the wrong foil shape in the impeller. This failure will be uncovered by the performance of the foil under the other testing. We don't want to purposely make a bad design to see if a bad design will cause a problem.

A quick word about managing the hypothesis matrix: I simply sorted the potential failures and mechanisms and pasted them as a column (failures) and a row (mechanisms) and then used the command "=CONCATENATE(B$2," will not cause ",$A3)" to build the text in the table. In this way, a small change to the DFMEA can be updated in the hypothesis matrix without a big hassle.

Next, we identify the tests that could evaluate each hypothesis in the matrix. Remember that some hypotheses will not need to be tested as we can assume that a biological attack (insects) will not cause abrasion. Again, a section of the full spreadsheet is shown here.

Table 14-3:

	Biological	chemical attack	current back through armature	fatigue - wire motion	impact	inductance / back emf /poor phasing with house AC	no stiffeners	poor angular control - non parallel	sharp objects	thermal cycle	thermal cycling	torque	unknown - improper sizing of standard pipe sizes	UV
abrasion -wear corrosion - loss of structural integrity - impeller binding - power loss poly coating	Biological will not cause abrasion -wear corrosion - loss of structural integrity - impeller binding - power loss poly coating	chemical attack will not cause abrasion -wear corrosion - loss of structural integrity - impeller binding - power loss poly coating	current back through armature will not cause abrasion -wear corrosion - loss of structural integrity - impeller binding - power loss poly coating	fatigue - wire motion will not cause abrasion -wear corrosion - loss of structural integrity - impeller binding - power loss poly coating	impact willnot cause abrasion -wear corrosion - loss of structural integrity - impeller binding - power loss poly coating	inductance/ back emf /poor phasing with house AC will not cause abrasion -wear corrosion - loss of structural integrity - impeller binding - power loss poly coating	no stiffeners will not cause abrasion -wear corrosion - loss of structural integrity - impeller binding - power loss poly coating	poor angular control - non parallel will not cause abrasion -wear corrosion - loss of structural integrity - impeller binding - power loss poly coating	sharp objects will not cause abrasion -wear corrosion - loss of structural integrity - impeller binding - power loss poly coating	thermal cycle will not cause abrasion -wear corrosion - loss of structural integrity - impeller binding - power loss poly coating	thermal cycling will not cause abrasion -wear corrosion - loss of structural integrity - impeller binding - power loss poly coating	torque will not cause abrasion -wear corrosion - loss of structural integrity - impeller binding - power loss poly coating	unknown - improper sizing of standard pipe sizes will not cause abrasion -wear corrosion - loss of structural integrity - impeller binding - power loss poly coating	UV will not cause abrasion -wear corrosion - loss of structural integrity - impeller binding - power loss poly coating
arcing voltage spike - transformer damage - fire wire terminal	Biological will not cause arcing voltage spike - transformer damage - fire wire terminal	chemical attack will not cause arcing voltage spike - transformer damage - fire wire terminal	current back through armature will not cause arcing voltage spike - transformer damage - fire wire terminal	fatigue - wire motion will not cause arcing voltage spike - transformer damage - fire wire terminal	impact will not cause arcing voltage spike - transformer damage - fire wire terminal	inductance back emf /poor phasing with house AC will not cause arcing voltage spike - transformer damage - fire wire terminal	no stiffeners will not cause arcing voltage spike - transformer damage - fire wire terminal	poor angular control - non parallel will not cause arcing voltage spike - transformer damage - fire wire terminal	sharp objects will not cause arcing voltage spike - transformer damage - fire wire terminal	thermal cycle will not cause arcing voltage spike - transformer damage - fire wire terminal	thermal cycling will not cause arcing voltage spike - transformer damage - fire wire terminal	torque will not cause arcing voltage spike - transformer damage - fire wire terminal	unknown - improper sizing of standard pipe sizes will not cause arcing voltage spike - transformer damage - fire wire terminal	UV will not cause arcing voltage spike - transformer damage - fire wire terminal
arcing voltage spike - transformer damage - fire AC terminal	Biological will not cause arcing voltage spike - transformer damage - fire AC terminal	chemical attack will not cause arcing voltage spike - transformer damage - fire AC terminal	current back through armature will not cause arcing voltage spike - transformer damage - fire AC terminal	fatigue - wire motion will not cause arcing voltage spike - transformer damage - fire AC terminal	impact will not cause arcing voltage spike - transformer damage - fire AC terminal	inductance /back emf /poor phasing with house AC will not cause arcing voltage spike - transformer damage - fire AC terminal	no stiffeners will not cause arcing voltage spike - transformer damage - fire AC terminal	poor angular control - non parallel will not cause arcing voltage spike - transformer damage - fire AC terminal	sharp objects will not cause arcing voltage spike - transformer damage - fire AC terminal	thermal cycle will not cause arcing voltage spike - transformer damage - fire AC terminal	thermal cycling will not cause arcing voltage spike - transformer damage - fire AC terminal	torque will not cause arcing voltage spike - transformer damage - fire AC terminal	unknown - improper sizing of standard pipe sizes will not cause arcing voltage spike - transformer damage - fire AC terminal	UV will not cause arcing voltage spike - transformer damage - fire AC terminal
arcing voltage spike - transformer damage - fire wire terminal	Biological will not cause arcing voltage spike - transformer damage - fire wire terminal	chemical attack will not cause arcing voltage spike - transformer damage - fire wire terminal	current back through armature will not cause arcing voltage spike - transformer damage - fire wire terminal	fatigue - wire motion will not cause arcing voltage spike - transformer damage - fire wire terminal	impact will not cause arcing voltage spike - transformer damage - fire wire terminal	inductance / back emf /poor phasing with house AC will not cause arcing voltage spike - transformer damage - fire wire terminal	no stiffeners will not cause arcing voltage spike - transformer damage - fire wire terminal	poor angular control - non parallel will not cause arcing voltage spike - transformer damage - fire wire terminal	sharp objects will not cause arcing voltage spike - transformer damage - fire wire terminal	thermal cycle will not cause arcing voltage spike - transformer damage - fire wire terminal	thermal cycling will not cause arcing voltage spike - transformer damage - fire wire terminal	torque will not cause arcing voltage spike - transformer damage - fire wire terminal	unknown - improper sizing of standard pipe sizes will not cause arcing voltage spike - transformer damage - fire wire terminal	UV will not cause arcing voltage spike - transformer damage - fire wire terminal

	ingress of moisture - chemicals	installation	leaves - debris in pipe	mechanical fatigue	Moisture	motion of armature	motion of case	motion of magnet	vibration	water	water flow - fatigue	wire flexes at grommet
abrasion -wear corrosion - loss of structural integrity - impeller binding - power loss poly coating	ingress of moisture - chemicals will not cause abrasion -wear corrosion - loss of structural integrity - impeller binding - power loss poly coating	installation will not cause abrasion -wear corrosion - loss of structural integrity - impeller binding - power loss poly coating	leaves - debris in pipe will not cause abrasion -wear corrosion - loss of structural integrity - impeller binding - power loss poly coating	mechanical fatigue will not cause abrasion -wear corrosion - loss of structural integrity - impeller binding - power loss poly coating	Moisture will not cause abrasion -wear corrosion - loss of structural integrity - impeller binding - power loss poly coating	motion of armature will not cause abrasion -wear corrosion - loss of structural integrity - impeller binding - power loss poly coating	motion of case will not cause abrasion -wear corrosion - loss of structural integrity - impeller binding - power loss poly coating	motion of magnet will not cause abrasion -wear corrosion - loss of structural integrity - impeller binding - power loss poly coating	vibration will not cause abrasion -wear corrosion - loss of structural integrity - impeller binding - power loss poly coating	water will not cause abrasion -wear corrosion - loss of structural integrity - impeller binding - power loss poly coating	water flow - fatigue will not cause abrasion -wear corrosion - loss of structural integrity - impeller binding - power loss poly coating	wire flexes at grommet will not cause abrasion -wear corrosion - loss of structural integrity - impeller binding - power loss poly coating
arcing voltage spike - transformer damage - fire wire terminal	ingress of moisture - chemicals will not cause arcing voltage spike - transformer damage - fire wire terminal	installation will not cause arcing voltage spike - transformer damage - fire wire terminal	leaves - debris in pipe will not cause arcing voltage spike - transformer damage - fire wire terminal	mechanical fatigue will not cause arcing voltage spike - transformer damage - fire wire terminal	Moisture will not cause arcing voltage spike - transformer damage - fire wire terminal	motion of armature will not cause arcing voltage spike - transformer damage - fire wire terminal	motion of case will not cause arcing voltage spike - transformer damage - fire wire terminal	motion of magnet will not cause arcing voltage spike - transformer damage - fire wire terminal	vibration will not cause arcing voltage spike - transformer damage - fire wire terminal	water will not cause arcing voltage spike - transformer damage - fire wire terminal	water flow - fatigue will not cause arcing voltage spike - transformer damage - fire wire terminal	wire flexes at grommet will not cause arcing voltage spike - transformer damage - fire wire terminal
arcing voltage spike - transformer damage - fire AC terminal	ingress of moisture - chemicals will not cause arcing voltage spike - transformer damage - fire AC terminal	installation will not cause arcing voltage spike - transformer damage - fire AC terminal	leaves - debris in pipe will not cause arcing voltage spike - transformer damage - fire AC terminal	mechanical fatigue will not cause arcing voltage spike - transformer damage - fire AC terminal	Moisture will not cause arcing voltage spike - transformer damage - fire AC terminal	motion of armature will not cause arcing voltage spike - transformer damage - fire AC terminal	motion of case will not cause arcing voltage spike - transformer damage - fire AC terminal	motion of magnet will not cause arcing voltage spike - transformer damage - fire AC terminal	vibration will not cause arcing voltage spike - transformer damage - fire AC terminal	water will not cause arcing voltage spike - transformer damage - fire AC terminal	water flow - fatigue will not cause arcing voltage spike - transformer damage - fire AC terminal	wire flexes at grommet will not cause arcing voltage spike - transformer damage - fire AC terminal
arcing voltage spike - transformer damage - fire wire terminal	ingress of moisture - chemicals will not cause arcing voltage spike - transformer damage - fire wire terminal	installation will not cause arcing voltage spike - transformer damage - fire wire terminal	leaves - debris in pipe will not cause arcing voltage spike - transformer damage - fire wire terminal	mechanical fatigue will not cause arcing voltage spike - transformer damage - fire wire terminal	Moisture will not cause arcing voltage spike - transformer damage - fire wire terminal	motion of armature will not cause arcing voltage spike - transformer damage - fire wire terminal	motion of case will not cause arcing voltage spike - transformer damage - fire wire terminal	motion of magnet will not cause arcing voltage spike - transformer damage - fire wire terminal	vibration will not cause arcing voltage spike - transformer damage - fire wire terminal	water will not cause arcing voltage spike - transformer damage - fire wire terminal	water flow - fatigue will not cause arcing voltage spike - transformer damage - fire wire terminal	wire flexes at grommet will not cause arcing voltage spike - transformer damage - fire wire terminal

Table 14-4:

For simplicity, I only put in the accelerated tests that would be used. We will add some performance and materials tests later.

With all the tests in, we need to rank the tests and failure modes for risk and cost so we can identify the sweet spot. Table 14-5 shows the ranking for risk and cost and then the sum of the two. The sweet spot is the minimum of the sum.

Table 14-5: Rank of each failure for risk of not knowing the failure exists at each level, the rank of cost for each test and the sum of cost and risk vs. development level.

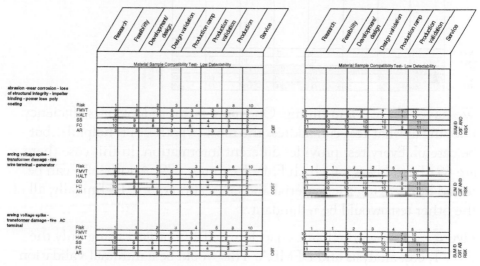

In addition to the risk is the timing required for each test vs. the time allotted for each stage of development. Table 14-6 shows the relative timing in days with the tests that cannot be run in the given phase of development highlighted.

Table 14-6: Test timing vs. time allotted for each stage of development. All numbers are in days. Note that some tests like Accelerated Reliability (AR) are specific to the failure mode being found and their timing may change with different failures. The CD-ROM shows the full spreadsheet for all potential failures from the DSG.

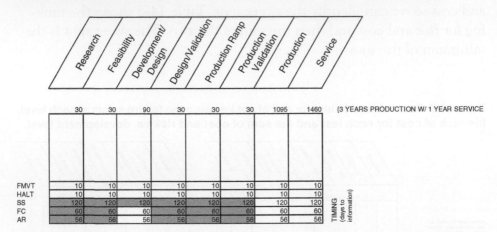

	Research	Feasibility	Development/ Design	Design/Validation	Production Ramp	Production Validation	Production	Service	
	30	30	90	30	30	30	1095	1460	(3 YEARS PRODUCTION W/ 1 YEAR SERVICE
FMVT	10	10	10	10	10	10	10	10	
HALT	10	10	10	10	10	10	10	10	
SS	120	120	120	120	120	120	120	120	TIMING (days to information)
FC	60	60	60	60	60	60	60	60	
AR	56	56	56	56	56	56	56	56	

We have risk, cost and timing. One more factor to look at is efficiency. Looking at what test can address the most hypotheses is helpful...but be careful. Every test provides different information. In this case there are 691 hypotheses, of which FMVT can test 679. Another 12 can be tested by checking the material and fabrication costs. Technically, all of the other tests would be redundant.

Figure 14-3 shows the Gant chart for the DSG project using only the FMVT, cost analysis and DFMEA. This would be a minimal validation plan. Before we accept this, we should take a good look at the RISK that we would be incurring by relying on such a brief test plan.

Examine the following risks: timing and business model. We have to have a product that meets the cost requirements (we have that covered), and is reliable enough to last longer than the payback period. Currently, we have no way of estimating the time-to-failure—except the actual field experience (a little late). This is not to say that all of the hypotheses are not addressed, they're just not addressed in a definitive way. The FMVT will identify failure modes effectively, and the design maturity will quantify if the design is robust, but no test quantifies a minimum life or reliability.

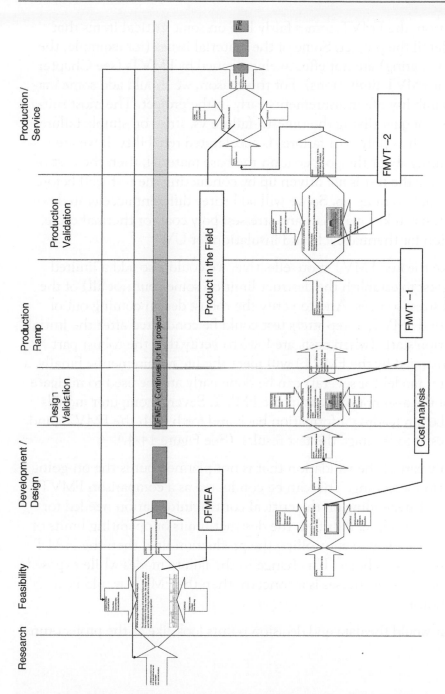

Figure 14-3: Minimal Gant chart with timing by phase. The Gant chart is on the CD-ROM and you will be able to zoom-in on the details.

In addition, the FMVT comes fairly late on some critical items that could derail the project. Some of the material issues (for example, the polymer coating) are not effectively screened by FMVTs (see Chapter 10 about FMVT limitations). For this reason, we should add some material durability life measurements early in the project. The most suitable test for quantifying the time-to-failure vs. stress on simple failure modes (such as poly-coat failure) is accelerated reliability. If we are only talking about the poly coat on the base material, then the cost of conducting the test is *not* driven up by conducting the test well before the production ramp up. So we will add three different accelerated reliability tests on key materials and stresses: poly coat for thermal cycle, insulation for thermal cycle and insulation for UV.

To make the first FMVT more effective, we could also add a limited HALT test to establish the destruct limit on some (but not all) of the critical stress sources. And to verify the robust design coming out of the second FMVT, a step stress test could be conducted after the initial production starts (when costs are low) to verify that the robust part (demonstrated by the FMVT) will meet the life requirements. Finally, a computer model (or several) can be done early and be used to mitigate some of the assumption risks in the FMVT. Several computer models would be run feeding information back and forth with the FMVT (and other tests) to leverage the test results. (See Figure 14-4.)

The only part of the validation that is not planned out is the on-going production screening. This can be conducted as a comparison FMVT or a HALT depending on the critical control information needed for the production line. If drift in the destruct limits or operating limits of the units were key to controlling the production line, then the HALT would be a good choice. If a change in the failure modes while exposed to a wide range of stresses is a concern, then the FMVT would be a good choice.

So what would the data and decision points look like as the project runs?

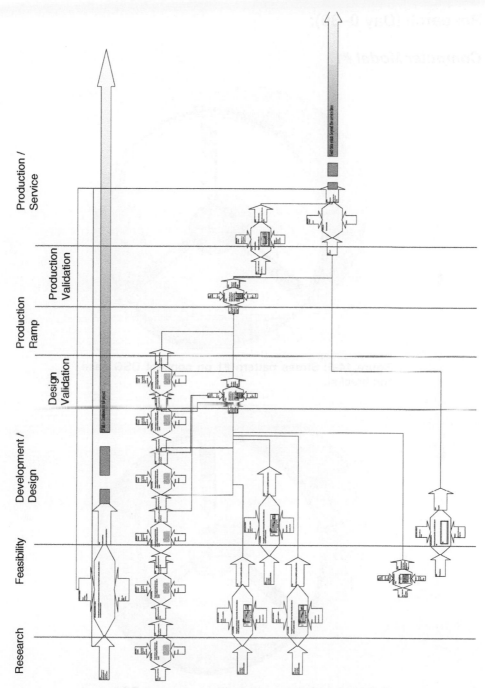

Figure 14-4: Full validation plan. See the CD-ROM for the full plan.

Research (Day 0–30):

Computer Model #1

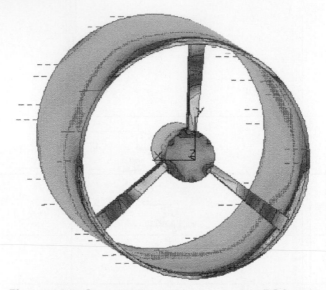

Figure 14-5: Stress pattern #1 on concept DSG case and bracket.

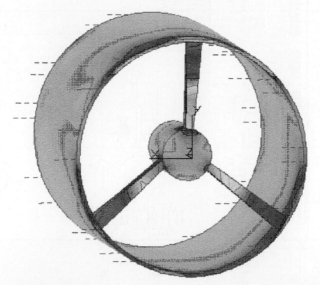

Figure 14-6: Stress pattern #2 on concept DSG case and bracket.

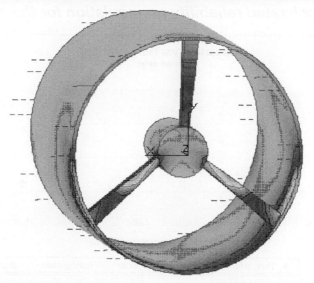

Figure 14-7: Stress pattern #3 on concept DSG case and bracket.

Start of accelerated reliability on poly coat in thermal cycle:

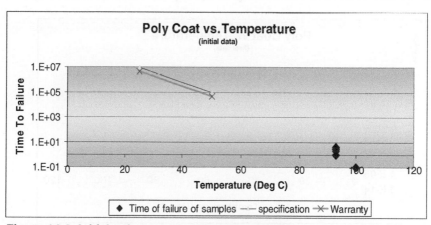

Figure 14-8: Initial poly coat vs. temperature data.

Start of accelerated reliability on insulation for thermal cycle:

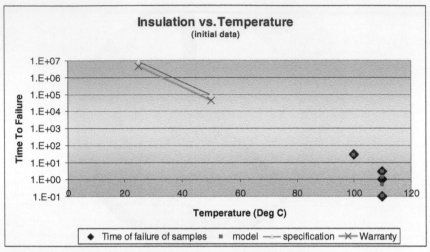

Figure 14-9: Initial data for insulation vs. temperature.

Feasibility (Day 30–60):

END of accelerated reliability on poly coat in thermal cycle:

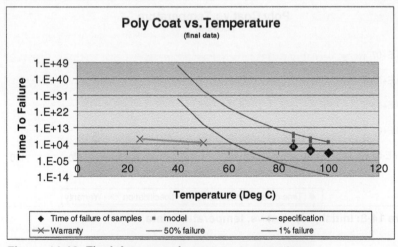

Figure 14-10: Final data on poly coat vs. temperature.

END of accelerated reliability on insulation for thermal cycle:

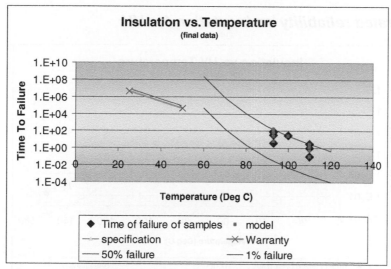

Figure 14-11: Final data on insulation vs. temperature

So far so good. Accelerated reliability data indicates that the poly coat and insulation shouldn't be a problem. The computer models show low stress.

HALT (thermal):

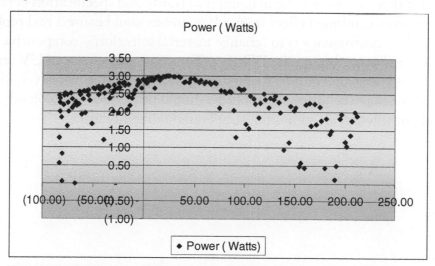

Figure 14-12: Results of HALT – power output as a function of temperature.

Development/Design (Day 60–150)

Accelerated reliability on insulation for UV:

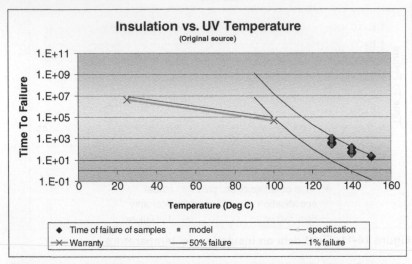

Figure 14-13: Insulation vs. UV temperature. Note that the 1% failure rate line is only marginally above the warranty and specification target.

There is a potential problem here: The accelerated reliability on the insulation vs. UV light indicates that the material is only marginally better than needed for the maximum warranty and specification conditions. Any combined effect from other stresses could cause a real problem. The contingency is to "change material selection – competitive sourcing," according to the DFMEA. The combined effects of UV and other stress sources will be found during the first FMVT. However, for good measure the alternative sources should be pursued immediately and put through an additional accelerated reliability test.

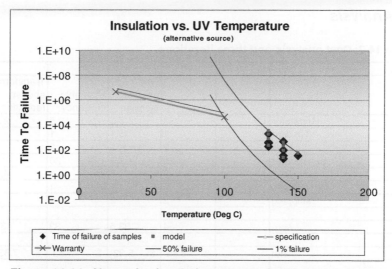

Figure 14-14: Alternative insulation source. Good news—average is better. Bad news—standard deviation is higher. This means that on average, the failure rate may actually be worse, even though the average product is better.

Oh, great. The alternative source is a mixed bag. On the one hand, the average time-to-failure is better. The slap in the face is that the 1% failure rate is worse due to a larger standard deviation. Better stick with the current source, but watch it closely during the FMVT. One other possible solution: use the alternative source, but find a way to "sort" the weaker material out.

Cost Analysis

Table 14-7: Cost analysis and three iterations.

price	$ 38.83	Base	new generator supplier	assembly efficiency	Transformer cost reduction
Generator		3	2.5	2.5	2.5
30 m wire		1	1	1	1
shell		2	2	2	2
transformer		20	20	20	18
assembly		5	5	4.5	4.5
packaging		1	1	1	1
shipping		2	2	2	2
sum		34	33.5	33	31
net		$ 4.83	$ 5.33	$ 5.83	$ 7.83
gross margin		12%	14%	15%	20%
overhead (on a million units)		$ 5.00	$ 5.00	$ 5.00	$ 5.00
Profit/ (loss)		$ (0.17)	$ 0.33	$ 0.83	$ 2.83

Costs look reasonable, provided the cheaper parts work and management can keep to $5.00 overhead per part.

Design Validation (Day 150–180)

FMVT – 1

Table 14-8: Stress plan for FMVT.

Level	Biological {ml of crickets and wax worms}	chemical attack {% HCL in water}	current back through armature amps - 10 ms once/min	fatigue - wire motion mm of displacement	impact m drop of 50 mg ball	installation Installations per level	leaves - debris in pipe % leaves (maple) by mass in water
1.00	10.00	2.00	0.20	10.00	1.00	1.00	1.00
2.00	20.00	4.56	0.29	20.00	2.00	2.00	3.11
3.00	30.00	7.11	0.38	30.00	3.00	3.00	5.22
4.00	40.00	9.67	0.47	40.00	4.00	4.00	7.33
5.00	50.00	12.22	0.56	50.00	5.00	5.00	9.44
6.00	60.00	14.78	0.64	60.00	6.00	6.00	11.56
7.00	70.00	17.33	0.73	70.00	7.00	7.00	13.67
8.00	80.00	19.89	0.82	80.00	8.00	8.00	15.78
9.00	90.00	22.44	0.91	90.00	9.00	9.00	17.89
10.00	100.00	25.00	1.00	100.00	10.00	10.00	20.00
Notes	Packed in cracks of installation			200 mm from terminal	glass ball		

Level	mechanical fatigue {water provides this}	Moisture {water provides this}	sharp objects 10 mg hardened steel cube / per level	Temperature (Cold) deg C	Temperature (Hot) deg C	Ramp deg / min	torque {water provides this}
1.00	1.00	1.00	1.00	(15.00)	25.00	15.00	1.00
2.00	1.00	1.00	2.00	(20.00)	35.00	15.00	1.00
3.00	1.00	1.00	3.00	(25.00)	45.00	15.00	1.00
4.00	1.00	1.00	4.00	(30.00)	55.00	15.00	1.00
5.00	1.00	1.00	5.00	(35.00)	65.00	15.00	1.00
6.00	1.00	1.00	6.00	(40.00)	75.00	15.00	1.00
7.00	1.00	1.00	7.00	(45.00)	85.00	15.00	1.00
8.00	1.00	1.00	8.00	(50.00)	95.00	15.00	1.00
9.00	1.00	1.00	9.00	(55.00)	105.00	15.00	1.00
10.00	1.00	1.00	10.00	(60.00)	115.00	15.00	1.00
Notes	Mechanical fatigue comes from the water running						

Level	unknown - improper sizing of standard pipe sizes mm too small in circumfrence	UV delta Deg C	v ibration peak g	water Liter/sec	water flow - fatigue {water provides this}	wire flexes at grommet mm of displacement
1.00	2.00	10.00	6.00	3.00	10.00	10.00
2.00	4.00	20.00	12.00	6.00	20.00	20.00
3.00	6.00	30.00	18.00	9.00	30.00	30.00
4.00	8.00	40.00	24.00	12.00	40.00	40.00
5.00	10.00	50.00	30.00	15.00	50.00	50.00
6.00	12.00	60.00	36.00	18.00	60.00	60.00
7.00	14.00	70.00	42.00	21.00	70.00	70.00
8.00	16.00	80.00	48.00	24.00	80.00	80.00
9.00	18.00	90.00	54.00	27.00	90.00	90.00
10.00	20.00	100.00	60.00	30.00	100.00	100.00
Notes	nominal rectangular size reduced at each level	UV light creating a temperature gradient above ambient				200 mm from terminal

Table 14-9: FMVT – 1 incident log.

Event #	Observation	Time Under Test [min]	Incident #	Level
1	Start test	0	0	1
2	Power production reduced 3% during cold cycle	1	1	1
3	Power recovered during ramp up and hot	20	0	1
4	Power production reduced 3% during cold cycle		1	1
5	End of level 1	60	0	2
6	Start Level 2	60	0	2
7	Power production reduced 3% during cold cycle		1	2
8	Power recovered during ramp up and hot	80	0	2
9	"humming" coming from transformer - goes away at cold	90	2	2
10	Power production reduced 3% during cold cycle	10	1	2
11	end of level 2	120	0	2
12	Power production reduced 3% during cold cycle	10	1	3
13	Power production reduced 2% during cold cycle	30	1	3
14	Power recovered during ramp up and hot	140	0	3
15	Phase angle "jumped" between house and transformer - trans shut down - restarted in phase 2 minutes later	145	3	3
16	"humming" coming from transformer - goes away at cold	150	2	3
17	Power production reduced 4% during cold cycle	70	1	3
18	end of level 3	180	0	3
19	Start Level 4	180	0	4
20	Power production reduced 2% during cold cycle	90	1	4
21	Power recovered during ramp up and hot	200	0	4
22	Phase angle "jumped" between house and transformer - trans shut down - restarted in phase 2 minutes later	205	3	4
23	Top spinner loose- rubbing on tube - grease was found on shaft and impeller replaced	210	4	4
24		215		4
25	Power production reduced 4% during cold cycle	210	2	4
26				4
27	Power production reduced 4% during cold cycle	30	1	4
28	End Level 4	240	0	4
29	Start Level 5	240	0	5
30	Power production reduced 5% during cold cycle	2	1	5
31	Power recovered during ramp up and hot	240	0	5
32	Phase angle "jumped" between house and transformer - trans shut down - restarted in phase 2 minutes later	265	3	5
33	Generator bracket #1 tearing at shell	270	5	5
34		265	0	5
35	"humming" coming from transformer - goes away at cold	270	2	5
36		270	0	5
37	Power production reduced 4% during cold cycle	90	1	5
38	End Level 5	300	0	5
39	Start Level 6	300	0	6
40	Power production reduced 3% during cold cycle	90	1	6
41	Power recovered during ramp up and hot	300	0	6
42	Phase angle "jumped" between house and transformer - trans shut down - restarted in phase 2 minutes later	325	3	6
43	Generator bracket #1 tearing at shell	325	5	6
44	Generator bracket #2 tearing at shell	325	5	6
45		330	2	6
46	insulation cracking where it passes through shell	330	6	6
47	Power production reduced 6% during cold cycle	50	1	6
48			0	7
49	Start Level 7	360	0	7
50	Power production reduced 6% during cold cycle	50	1	7
51	Power recovered during ramp up and hot	360	0	7
52	Phase angle "jumped" between house and transformer - trans shut down - restarted in phase 2 minutes later	385	3	7
53	Generator bracket #1 tearing more at shell	385	5	7
54	Generator bracket #2 tearing more at shell	385	5	7
55	"humming" coming from transformer - goes away at cold	390	2	7
56	insulation cracking where it passes from shell	390	6	7
57	Power production reduced 9% during cold cycle	10	1	7
58	End Level 7	420	0	7
59	Start Level 8	420	0	8
60	Power production reduced 25% during cold cycle	10	1	8
61	Power recovered during ramp up and hot	420	0	8
62	Phase angle "jumped" between house and transformer - trans shut down - restarted in phase 2 minutes later	445	3	8
63	Generator bracket #1 tearing more at shell	445	5	8
64	Generator bracket #2 tearing more at shell	445	5	8
65	Generator bracket #3 tearing at shell	450	6	8
66	insulation cracking where it passes from shell - threads of multi-strand wire breaking	450	6	8
67	Power production reduced 35% during cold cycle	70	1	8
68	End Level 8	480	0	8
69	Start Level 9	480	0	9
70	Power production reduced 2% during cold cycle	70	1	9
71	Power recovered during ramp up and hot	480	0	9
72	Generator "sticking" periodicly - tear down shows the bearings had water engress and corrosion	85	7	9
73	With power below oscillating above and below 1 W transformer tries to phase but can't - high heat build up	490	8	9
74	Phase angle "jumped" between house and transformer - trans shut down - restarted in phase 2 minutes later	505	3	9
75	Generator bracket #1 tearing more at shell	505	5	9
76	Generator bracket #2 tearing more at shell	505	5	9
77	Generator bracket #3 tearing at shell	510	5	9

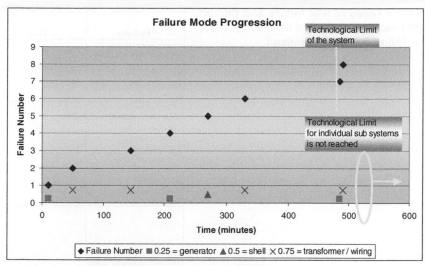

Figure 14-15: Failure progression from FMVT – 1.

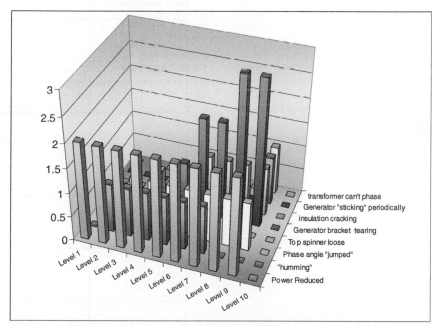

Figure 14-16: Histogram of failures from FMVT – 1. Notice that "power reduced" repeats often. This was predicted by the brief HALT done earlier.

Table 14-10: Times to first failure and design maturity calculations. Based on the numbers, failures 1–6 should be addressed.

Failure	Time
Power production reduced 3% during cold cycle	10
"humming" coming from transformer - goes away at cold	50
Phase angle "jumped" between house and transformer - trans shut down - restarted in phase 2 minutes later	145
Top spinner loose- rubbing on tube - grease was found on shaft - shaft and impeller replaced	210
Generator bracket #1 tearing at shell	270
insulation cracking where it passes from shell	330
Generator "sticking" periodically - teardown shows the bearings had water ingress and corrosion	485
With power below oscilating above and below 1 W transformer tries to phase but can't - high heat build up	490
DM	6.86
PDM1	1.47
PDM2	0.48
PDM3	0.33
PDM4	0.27
PDM5	0.24
PDM6	0.01

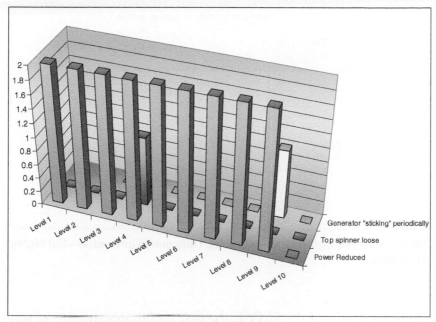

Figure 14-17: Histogram of generator failures.

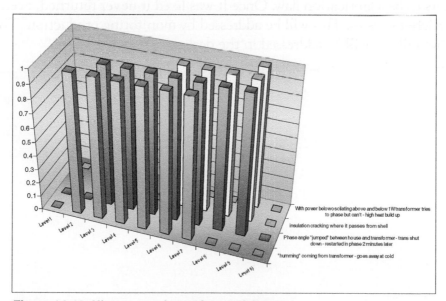

Figure 14-18: Histogram of transformer failures.

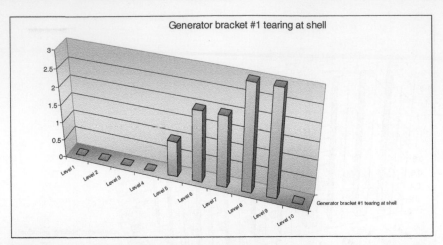

Figure 14-19: Histogram of shell failures. Only one failure mode—but highly repeatable.

Based on the results of the FMVT, the first six failures should definitely be addressed. The "top spinner loss," which was failure number four appears to be a fabrication flaw. Once it was fixed it never returned, even at higher stresses. This will be addressed by monitoring production. The other failures will be addressed in the design.

Production Validation (Day 180–210)

FMVT – 2

Table 14-11: Incident log from FMVT – 2. Notice that the product is now much improved.

Event #1	Observation	Time Under Test [min]	Incident #	Level
1	Start test	0	0	1
2				
3	Start of Level 2	60	0	2
4				
5	Start of Level 3	120	0	3
6	End of Level 3	180	0	3
7				
8	End of Level 4	240	0	4
9				
10	End of Level 5	300	0	5
11				
12	End of Level 6	360	0	6
13	Start of Level 7	360	0	7
14				
15	Start of Level 8	420	0	8
16				
17	Start of Level 9	480	0	9
18	Power production reduced 4% during cold cycle	520	1	9
19				
20	Start of Level 9	540	0	9
21	Power production reduced 3% during cold cycle	580	1	10
22	Phase angle "jumped" between house and transformer - trans shut down - restarted in phase 2 minutes later	585	2	10
23	Generator bracket #1 tearing at shell	599	3	10
24	End of Level 10	600	0	10

Table 14-12: Failure modes and Design Maturity (DM) from FMVT – 2. Notice that the DM is below 0.1.

Failure Number	Failure	Time
1	Power production reduced during cold cycle	520
2	Phase angle "jumped" between house and transformer - trans shut down - restarted in phase 2 minutes later	585
3	Generator bracket #1 tearing at shell	599
	DM	0.08
	PDM1	0.02

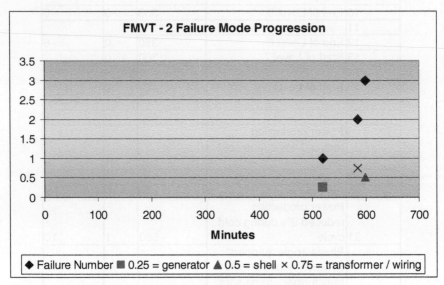

Figure 14-20: Failure mode progression from FMVT – 2.

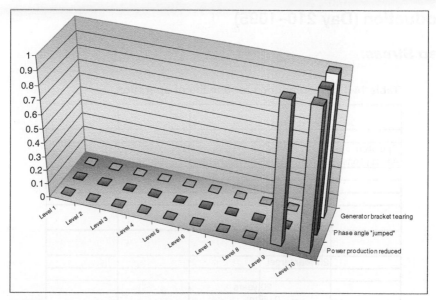

Figure 14-21: Histogram of failures from FMVT – 2.

Good, robust design. Ready for production.

Production (Day 210–1095)

Step Stress:

Table 14-13: Assumptions for one life of operation.

Defining One Life:				
Operation Assumptions				
	1 meter of rain fall per year in 50 mm - 2h spurts			
	5 year pay back			
	10 year warranty			
	20 year target			
	20	years		
	0.05	m		
	2	hours for 0.05 meters		
	20	times a year		
	400	in a life		
	2	hrs cycle "raining"		
	2	hrs "dry"		
	1600	hrs of operation		

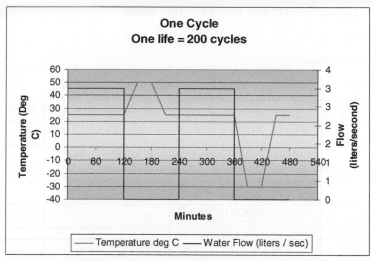

Figure 14-22: Step stress – life portion cycle. Two hundred cycles is assumed to equal one life.

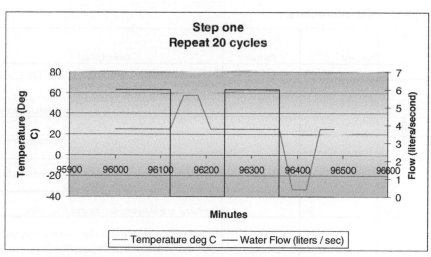

Figure 14-23: Step one cycle. Twenty cycles for one step.

Table 14-14: Life portion results and failures during steps.

12 parts passed to one life

90% Reliable with a 70% Confidence

Failures During Stepping

Step	Time [hrs]	Failure	Failure #
3	1942.5	Power production reduced 3% during cold cycle	1
4	2106.5	Phase angle "jumped" between house and transformer - trans shut down - restarted in phase 2 minutes later	2
4	2107	Phase angle "jumped" between house and transformer - trans shut down - restarted in phase 2 minutes later	2
4	2111	Power production reduced 4% during cold cycle	1
5	2279.5	Phase angle "jumped" between house and transformer - trans shut down - restarted in phase 2 minutes later	2
6	2442.5	Generator bracket #1 tearing at shell	3
7	2610.5	Generator bracket #1 tearing more at shell	3
7	2611	Generator bracket #2 tearing at shell	3
7	2614.5	Power production reduced 5% during cold cycle	1
7	2615.5	Generator bracket #3 tearing at shell	3

Table 14-15: Field return data.

Incident per hundred thousand	Percent	Failure	Description
1.76	9%	1	Power production reduced 3% during cold cycle
5.49	28%	2	Phase angle "jumped" between house and transformer - trans shut down - restarted in phase 2 minutes later
6.37	32%	3	Generator bracket #1 tearing at shell
3.30	17%	4	"humming" coming from transformer - goes away at cold
2.31	12%	5	Top spinner loose - rubbing on tube - grease was found on shaft - shaft and impeller replaced
0.55	3%	6	insulation cracking where it passes from shell
-	0%	7	Generator "sticking" periodically - tear down shows the bearings had water ingress and corrosion
0.11	1%	8	With power below oscilating above and below 1 W transformer tries to phase but can't - high heat build up

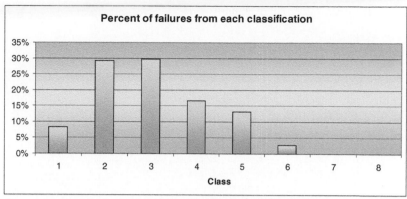

Figure 14-24: Percent of failures from each classification.

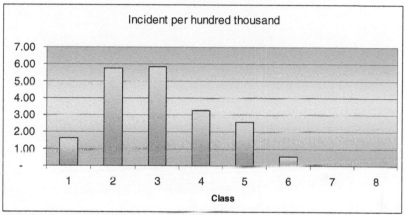

Figure 14-25: Incident rate. Graph shows how many of each class of failure per hundred thousand units produced.

All of the tables and figures from this example are on the accompanying CD-ROM. Many of them are Monte Carlo models. This just means that a math equation seeded with a random number was used to produce the data. Hitting F9 (or recalculating) will produce a new set of data.

Percent of failures from each classification

Figure 14-24: Percent of failures from each classification.

Incident per hundred thousand

Figure 14-25: Incident rate. Graph shows how many of each class of failure per hundred thousand units produced.

All of the tables and figures from this example are on the accompanying CD-ROM. Many of them are Monte Carlo models. This just means that a math equation seeded with a random number was used to produce the data. Hitting F9 (for recalculating) will produce a new set of data.

About the Author

Alexander (Alex) J. Porter is the Engineering Development Manager for Entela, Inc. and has been with the company since 1992. Since 1996, he has been developing accelerated testing methods for mechanical components and systems. Alex has three patents related to accelerated testing equipment and has published over thirty articles, technical papers and presentations on accelerated testing. Alex is chairing an SAE committee that is writing an Accelerated Testing Supply Chain Implementation Guide. His work in the past has included implementation of FEA in a laboratory setting and development of a thermal management system for an advanced data acquisition package developed by NASA's Drydon Flight Research facility. Alex is a member of SAE and IEEE. He holds a B.S. in Aircraft Engineering and an M.S. in Mechanical Engineering, both from Western Michigan University.

Index

A

accelerated reliability, 55, 111, 113, 180, 216, 219, 222

B

Best Practices committee, 31
boundary conditions, 172
business model, 9, 21, 40, 205
business structure, 14, 50, 206
business style, 109
business type, 52

C

commodity, 10
commonality, 26
computer model, 170, 216, 218, 221
computer modeling, 1, 27, 55, 180, 188
contingency, 82, 202, 222
continuous variable, 92
controls, 78
criticality, 78
cross-functional teams, 20

D

data, 9, 15
decisions, 1
design failure mode effects analysis (DFMEA), 55, 77, 140, 180, 186, 193, 208, 214, 222

design maturity (DM), 144, 164
design validation, 47, 80, 225
destruct limit, 107, 128, 142, 150, 216
detectability, 78
development, 222
development/design, 47
DFMEA, 55, 77, 140, 180, 186, 193, 208, 214, 222
DISCRETE variable, 93, 95

E

engineering analysis, 1

F

failure modes and effects analysis (FMEA), 1
failure mode verification testing (FMVT), 55, 139, 141, 180, 183, 186, 214, 216, 222
fault tree analysis (FTA), 1, 55
feasibility, 47, 220
feasibility tests, 27
field, 95, 157, 185
finite element analysis (FEA), 4
FMVT, 55, 139, 141, 180, 183, 186, 214, 216, 222
fully censored, 87, 101, 180, 186
functional item, 78
functional testing, 28

Printed and bound by CPI Group (UK) Ltd, Croydon, CR0 4YY

03/10/2024

01040432-0008